Immuno-biotechnology

By

Dr. P.R. Yadav

Lecturer

Department of Zoology

D.A.V. College

Muzaffarnagar (U.P.)

&

Dr. Rajiv Tyagi

Department of Zoology

M.M. College

Modi Nagar (U.P.)

DISCOVERY PUBLISHING HOUSE

NEW DELHI-110002

First Published-2005
Reprinted: 2013
ISBN 81-8356-039-3

Published by

DISCOVERY PUBLISHING HOUSE
4831/24, Ansari Road, Prahlad Street,
Darya Ganj, New Delhi-110002 (India)
Phone: 23279245 • Fax: 91-11-23253475
E-mail:dphtemp@indiatimes.com

Printed at:
: Dynamic printers,

Preface

The present title "Immuno-biotechnology" is one of the fastest emerging field because of its importance in areas like biomedical engineering, pharmaceutical, industry and biochemistry. It stands out in the order of other sciences and technology because of the single reason that every development in this field has a direct impact on man. Immuno-biotechnology provides a concise yet comprehensive introduction to the various basic techniques used in the related fields. The presentation of the text is simple and systematic. The basic concepts have been clearly explained and their functions are adequately highlighted. Recent developments have also been included to provide a complementary understanding of the subject.

In the preparation of this book large number of books and research papers have been consulted. So no authenticity is claimed.

A text book can not be written without the support and professional contributions of many people. This book is no exceptional. The author is grateful to those teachers and colleagues whose stimulating discussions have clarified certain vague points for him, but all errors, omissions, and corrections needed are solely his responsibility.

The author tried hard to be accurate and upto date in statement and realises the impossibility of completely avoiding errors therefore, the author will greatly appreciate having his attention called to any questionable statements.

The author is deeply grateful to for his expertise and advice during the preparation of the present title.

The author expresses his gratitude to Mr. Wasan and staff of M/s Discovery Publishing House for their whole hearted co-operation in the publication of this book.

Authors

Preface

CONTENTS

1

TOOLS USED IN IMMUNO-BIOTECHNOLOGY

Immunoassays are tests utilizing interactions between antibodies and antigens, and can be used to detect and quantitate either an antigen or an antibody. These assays are powerful tools used widely in medicine, agriculture, and food services, and in many fields of basic research. Three general classes of immunoassays and some common variations are described briefly below.

Antibody capture assays can be used to detect and quantitate either antigens or antibodies. The general protocol involves immobilizing an antigen on a solid phase (e.g. a well of a plastic dish or a membrane), and then adding antibody and allowing it to bind to the immobilized antigen. The solid phase is then washed to remove any unbound material. To detect antibody binding, the antibody can be labeled directly, or a labeled secondary reagent, which will specifically recognize the antibody, can be used (methods of labeling are discussed below). The amount of antibody that is bound determines the strength of the signal.

Antigen capture assays are used primarily to detect and quantitate antigens in a sample. Unlabeled antibody is immobilized on the solid phase, and labeled antigen is allowed to bind to the immobilized antibody. The most common variations on this method involve competition between a known quantity of labeled antigen and unlabeled antigen in a test sample. When these two samples are mixed and added to the immobilized antibody, they will compete for binding to the antibody binding sites. If the test sample contains a high concentration of the antigen, it will compete effectively with the labeled antigen and little labeled antigen will bind. Thus, the amount of label present (i.e. the signal detected) will be inversely proportional to the amount of antigen present in the test sample.

Two antibody (*sandwich*) assays are used primarily to determine the antigen concentration in unknown samples; the assays are quick and accurate. The assay requires two antibodies that bind to non-overlapping regions of the same antigen. One antibody is immobilized to the solid phase, and the antigen in the test sample is allowed to bind. Unbound material is washed away, and a labeled, second antibody is added which will bind to any antigen which bound to the immobilized antibody. The major advantages of this technique are that it is very sensitive and specific, and crude antigen samples can be used.

All immunoassays rely on labeled antigens, antibodies, or secondary reagents for detection. These proteins can be labeled with radioactive molecules (usually radioactive iodine), enzymes with colour-producing substrates, or fluorescent dyes. Of these, radioactive labeling can be used for almost all types of immunoassay; it is easy to detect and quantitate. However, it is not always desirable to use radioactive materials due to safety and disposal considerations. Enzyme-labeled reagents yield semi-quantitative results and are now very commonly used, particularly in clinical labs and other uses outside of research labs. A colourless substrate for the enzyme, which turns colour when acted upon by the enzyme, is used for detection. An immunoassay using an enzyme label is usually called an ELISA (Enzyme-Linked Immunosorbent Assay).

This module will examine immunoassays which are used commercially and which are readily available as kits. All utilize enzyme labels. The immunoassays will be performed according to the instructions provided with the kit. Using the information provided in this introduction and in the kit instructions, the type of immunoassay represented by the kit, and the nature of the interaction between antibodies and antigens in the assay will be determined.

Safety Guidelines

Standard lab safety procedures should be followed. Note any special safety precautions described in any kit instructions used.

No special precautions are required, unless indicated for a particular kit or if human biological samples are used. The latter should be disposed of as biohazard waste.

Experimental Outline

Timetable of events

Most kits are designed to be performed quickly, and generally take 10-30 minutes to complete. This lab can be included in another related lab either to illustrate applications of antibodies to start the lab or during an incubation or other waiting period. Alternatively, several kits can be performed and analyzed to fill a 2- or 3.hour lab period.

Perform test; follow test instructions - generally 10-30 min.

Analyze results. Interpret test results as positive or negative and determine location of antibodies and antigens, mechanism of labeling - about 15 min.

Materials

Immunoassay kit and instructions

Test sample and/or control samples, as Indicated by instructor

Antibody based test kits, such as home pregnancy kits, are available in drug stores. These kits illustrate how simplified the use of antibodies as tools has become. Many other types of kits are available commercially.

Test samples and controls. Control urine samples (containing high and low levels of human chorionic gonadotrophin) are available commercially for pregnancy test kits. These purchased controls are usually preferable to obtaining urine samples from volunteers.

Most drug stores, grocery stores, and discount stores carry a large selection of home pregnancy test kits. One type that is easy to explain is the brand First Response, which is set up as described in the answer to question 1 above, with all of the components added in separate steps. Many other brands are set up with all reagents in a single test unit, and the urine is drawn by capillary action past all of the reagents. Other sources of test kits:

Fisher Scientific

711 Forbes Avenue, Pittsburgh, PA 15219; 214-562-8300

Carries control hCG urine samples for pregnancy tests, and a variety of clinical test kits.

BioMetallics, Inc.

Princeton, NJ; 1-800-999-1961

TARGET milk progesterone test kit

If possible, obtain fresh milk samples from a local farm. Milk from the grocery store yields a very low or negative results; this milk is usually obtained from a farm bulk tank so it tends to have an average level of progesterone, but apparently some antigenic activity is lost when the milk is homogenized. If you are in a rural area, a local farm supply store may carry similar test kits.

Agri-Diagnostics Assoc.

One Executive Drive, Moorestown, NJ 08057

REVEAL turf grass disease detection kits for identifying specific fungal pathogens; kits come with positive control samples, and test samples can be obtained from local golf courses. Be sure kits and control samples have not reached expiration dates.

Pre-lab Preparation

Generally none, other than obtaining kits and samples

Method

Perform the test using the samples provided, following the directions provided with the kit. Record the results of the test.

Results

Interpret the results of the test, considering any control sample run. What valid conclusion can be drawn from the test?

2

PRACTICAL IMMUNO-BIOTECHNOLOGY

ANTIBODIES AS RESEARCH AND DIAGNOSTIC TOOLS

Antibodies and Assays

Methods for measuring antigen-antibody reactions have been well established and induce those that have direct biologic relevance. The combination of Ab with biologically active Ag (virus, toxin, enzyme and hormone) can be detected by neutralization of the virus infection, toxicity, enzymatic and hormonal activity, respectively. Precipitation and agglutination have also been adapted for development of several useful assays. A variety of other assays have been developed which provide specific qualitative and quantitative measurement of Ag or Ab for both research and diagnostic purpose.

Table 2.1. Effects of combination of antigen and antibody.

Agglutination	Antigenic particle + specific Ab results in aggregation of particles
Precipitation	Soluble Ag + specific Ab results in lattice formation and precipitation
C Activation	Ag in solution or on particle + specific Ab results in activation of C
Cytolysis	Cell + anti-cell Ab + C may result in lysis of the cell
Opsonization	Antigenic particle + Ab + C enhances phagocytosis by Mo, MØ, PMNs
Neutralization	Toxins, viruses, enzymes, etc. + specific Abs may result in their inactivation

C—Complement; Mo—monocytes; MØ—macrophages; Ab—antibody; PMNs—polymorphonuclear cells.

Since the immune system recognizes and remembers virtually all Ags that are introduced into an individual, assays which demonstrate the presence of Ab to an organism in the serum of a patient have become a standard way of determining that the patient has had contact with, was infected by, the organism (e.g. the presence of Ab to HIV in the serum of a patient usually means that the patient has been infected with HIV). Alternatively, Abs with defined specificity (e.g. to Ags associated with cancer cells) can be used to determine the presence of disease associated Ags in a patient. Abs are also extremely important tools in molecular and cellular research as they permit the localization and characterization of Ags.

Precipitation and Agglutination

Precipitation Assays

As previously described, when there is both sufficient Ag and sufficient Ab, the combination of Ag and Ab proceeds until large aggregates are formed which are insoluble in water and precipitate (equivalence). The extent to which a lattice forms depends on the relative amounts of Ag and Ab present. Lattice formation, and precipitation are the basis for several qualitative and quantitative assays for Ag or Ab. These assays are done in semisolid gels into which holes are cut for Ag and/or for Ab and diffusion occurs until Ag and Ab are in equivalence and precipitate.

In *radial immunodiffusion*, Ab (e.g. horse anti-human IgG) is incorporated into the gel and Ag (e.g. human serum) is placed in a hole cut in the gel. Ag diffuses radially out of the well into the gel and interacts with the Ab forming a ring of precipitation, the diameter

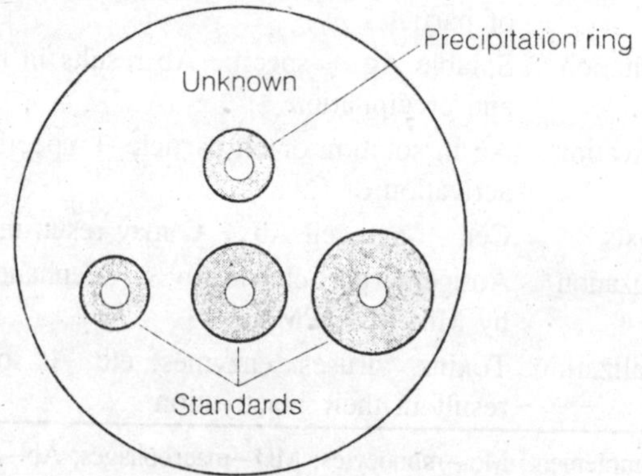

Fig. 2.1. Measurement of Ag by precipitation in gels.

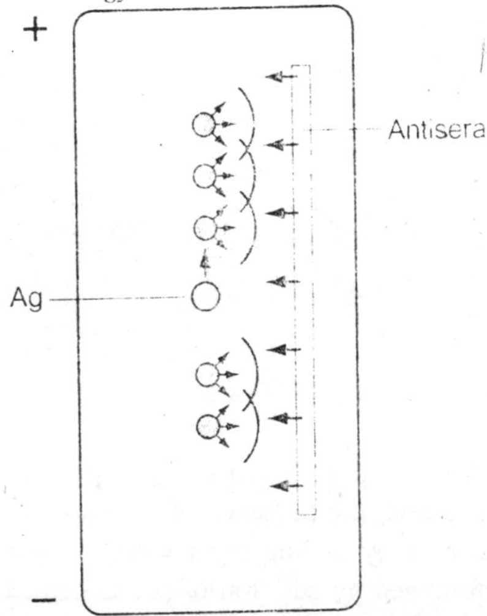

Fig. 2.2. Identification of antigens using gel electrophoresis.

of which is related to the concentration of the Ag. Similar assays have been developed in which a voltage gradient (electrophoresis) is used to speed up movement of Ag into the Ab containing gel (rocket immunoelectrophoresis). In *immunoelectrophoresis*, Ags (e.g. serum) are placed in a well cut in a gel (without (Ab) and electrophoresed, after which a trough is cut in the gel into which Abs (e.g. horse anti-human) are placed. The Abs diffuse laterally to meet diffusing Ag, and lattice formation and precipitation occur permitting determination of the nature of the Ags.

Agglutination Assays

Agglutination involves the interaction of surface Ags on *insoluble particles* (e.g. cells) and specific Ab to these Ags. Ab thus links together (agglutinates) insoluble particles. Much smaller amounts of Ab suffice to produce agglutination than are needed for precipitation. For this reason, agglutination rather than precipitation may be used to determine blood types or if Ab to bacteria is present in blood as in indication of infection with these bacteria. Since IgM has 10 binding sites, whereas IgG has two, IgM is much more efficient at agglutinating particles or cells. Although Abs are frequently used by themselves to assay for the presence of an Ag, a second Ab is sometimes used in what is known as a *Coomb's test*.

In some instances, such as when an autoantibody has been produced against a given cell type, the cells will have human Ab bonded to them, and thus can be identified by a second Ab (an Ab to human

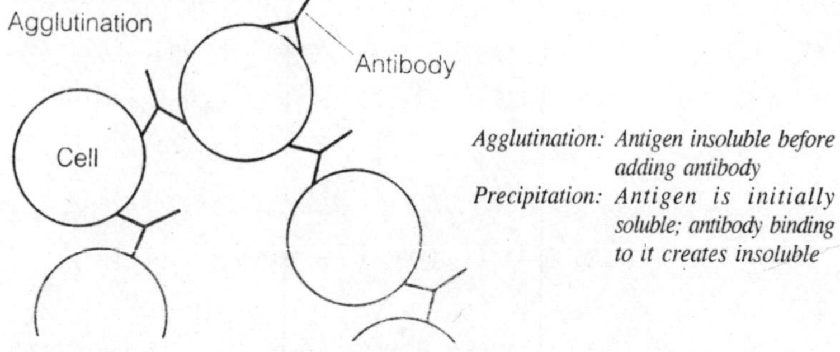

Fig. 2.3. Agglutination.

immunoglobulin) which will cause agglutination of the cells. In an *indirect Coomb's test*, the presence of circulating Ab to a cell surface Ag is demonstrated by adding the patient's serum to test cells (e.g. erythrocytes) followed by addition of Ab to human Ab.

Immunoassay

ELISA, RIA

The presence of Ab to a particular Ag in the serum of a patient can be determined using very sensitive radioimmunoassays (RIA) or enzyme-linked immuno-absorbent assays (ELISA). Such assays are of particular value in demonstrating Ab to Ags of infectious agents, e.g. virus, bacteria, etc. The presence of an Ab of particular isotype can also be determined using a modification of these assays. The radioallergosorbent test (RAST) uses as detecting ligand a radiolabeled Ab to human IgE and permits the measurement of specific IgE Ab to an allergen. ELISA and RIA also provide very specific and sensitive measurement of toxins, drugs, hormones, pesticides, etc., not only in serum, but also in water, foods and other consumer products. Based on these procedures, assays for nearly any Ag or Ab can be readily developed.

Immunofluorescence and Flow Cytometry

Although it is possible to use ELISA and RIA to evaluate the presence of an Ag on a cell, this is usually more conveniently done using Abs to which a fluorescent marker has been covalently attached. Moreover, in most cases a mAb is used and thus is highly specific for a particular molecule and a particular epitope on that molecule. This type of assay can be done using an Ab to the Ag which is directly fluorescent labeled (*direct* immunofluorescence) or by first incubating the unlabeled Ab with the cells (e.g. a mouse mAb to human T cells)

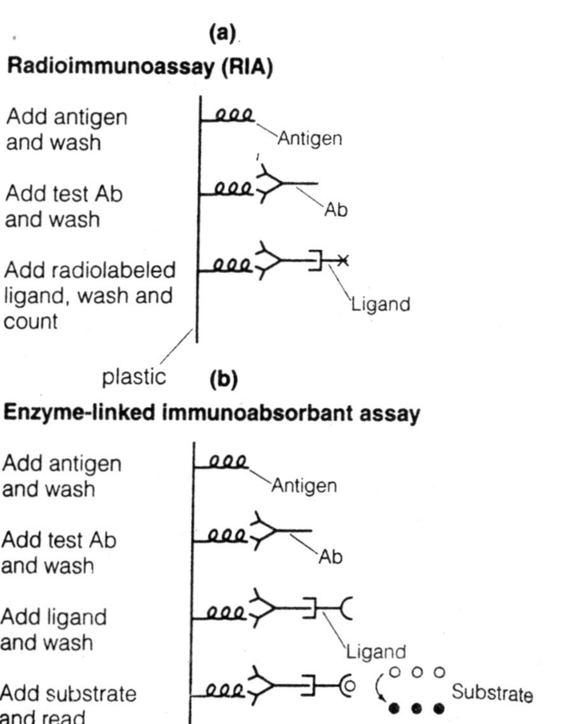

Fig. 2.4. (a) Radioimmunoassay (RIA); (b) Enzyme linked immunosorbent assay (ELISA).

and then, after washing away unbound Ab, adding a second fluorescent labeled Ab that reacts with the first Ab (e.g. a goat Ab to mouse immunoglobulin). This *indirect* immunofluorescent assay has two advantages, it has higher sensitivity and requires labeling of only one Ab, second Ab, because, in the example given, it can detect (react with) any mouse Ab. Fluorescent Abs to cell surface molecules (e.g. those which are tumor associated) are very useful in examining tissue sections for cells expressing the Ag. This assay is done by incubating the tissue section with the labeled Ab (for direct Immunofluorescence (IF) or unlabeled Ab, followed by labeled second Ab and then examining the tissue section using a fluorescent microscope.

These microscopes irradiate the tissue with a wavelength of light that excites the fluorescent label on the Ab to emit light at a different wavelength. This emitted light can be directly visualized, photographed and even quantitated. Moreover, it is possible to analyze a tissue sample using several different Abs at the same time, as each Ab a

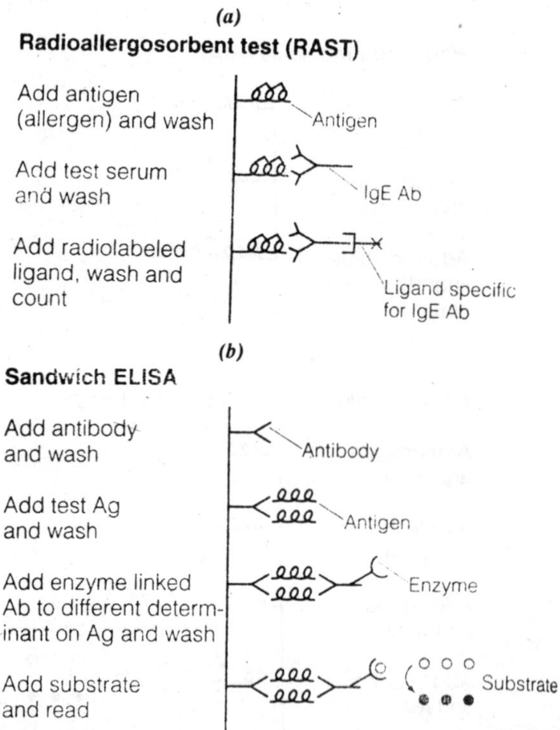

Fig. 2.5. (a) Radio allergosorbent test (RAST); (b)Sandwich ELISA and RIA.

wavelength distinct from the other. It is also possible to look for intracellular molecules (e.g. Abs) by first permeabilizing the cells and then doing the staining and fluorescence microscopy. Thus, one can

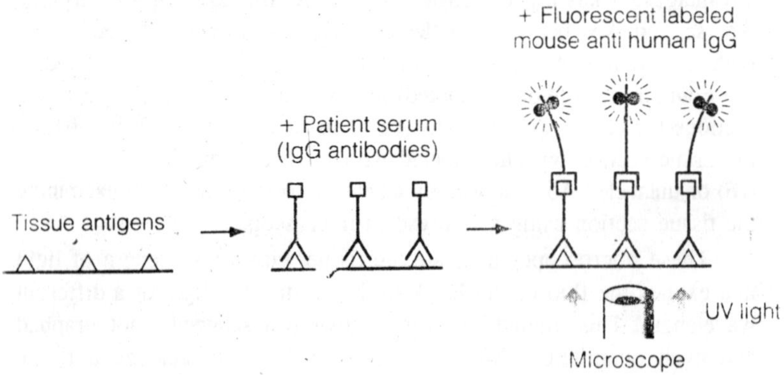

+ Fluorescent labeled mouse anti human IgG

+ Patient serum (IgG antibodies)

Tissue antigens

UV light

Microscope

Fig. 2.6. Indirect immunofluorescence assay.

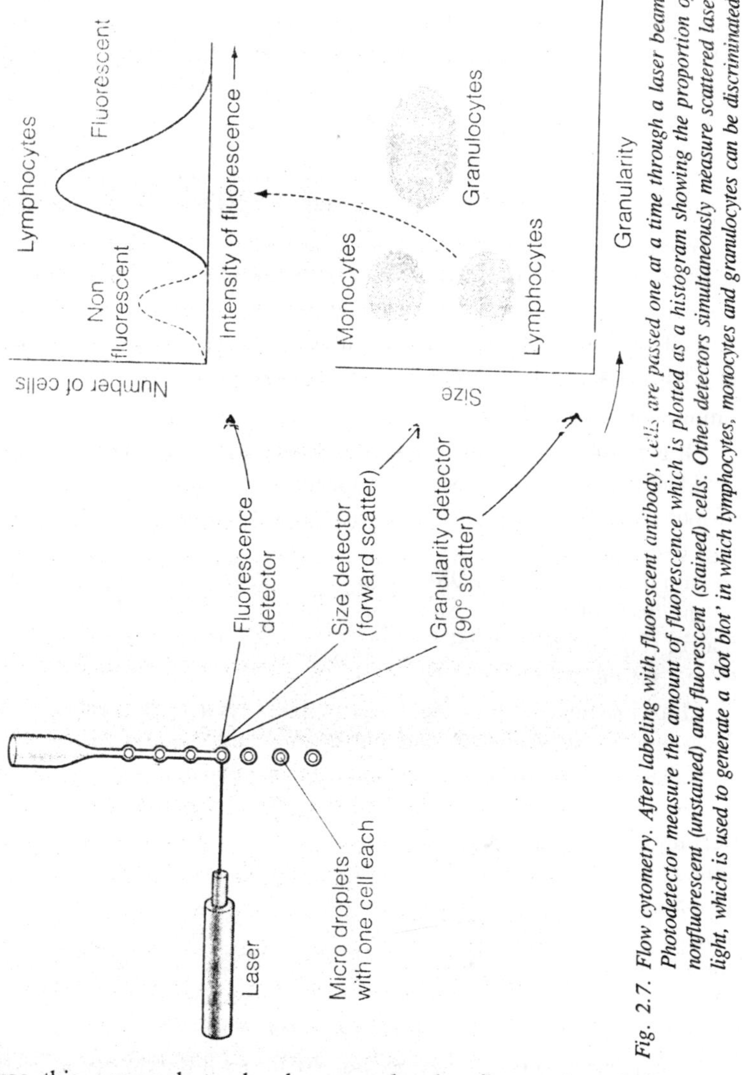

Fig. 2.7. Flow cytometry. After labeling with fluorescent antibody, cells are passed one at a time through a laser beam. Photodetector measure the amount of fluorescence which is plotted as a histogram showing the proportion of nonfluorescent (unstained) and fluorescent (stained) cells. Other detectors simultaneously measure scattered laser light, which is used to generate a 'dot blot' in which lymphocytes, monocytes and granulocytes can be discriminated.

use this approach to develop a molecular fingerprint of the cells associated with a tissue. Although fluorescence microscopy can be, and is, applied to the analysis of ***single cell suspensions***, another rather technologically sophisticated approach, ***flow cytometry***, is most often used. This assay uses the same basic staining procedures as described for fluorescence microscopy, followed by automated quantitation of the amount of fluorescence associated with individual cells.

In particular, the suspension of stained cells is fed to the flow cytometer which disperses the cells so they then pass single file through a focused laser beam which excites any fluorescent label associated with the cells. Those stained by the fluorescent Ab emit light that is detected and quantitated by optical sensors and the intensity of fluorescence is plotted in histograph form by a computer. This machine can analyze 1000 cells per second and provide quantitative data on the number of molecules of a particular kind on each cell. It can also analyze mixtures of cells and provide data on their size and granularity in addition to their expression of specific molecules. Some versions of this machine (*fluorescence activated cell sorter*) are also able to separate out cells into microdroplets and sort those expressing a selected amount of a particular Ag into a separate tube for further analysis or culture.

Immunoblotting

It is possible to combine various separation and detection procedures for identification and analysis of Ags and for evaluating the expression of molecules by single cells. Western blot analysis involves separating Ags by polyacrylamide gel electrophoresis (PAGE) in the presence of sodium dodecyl sulfate (SDS) which results in separation of molecules on the basis of size. These molecules are then transferred to another matrix (e.g. nitrocellulose) to form a pattern on the matrix identical to that on the gel.

Enzyme linked Ab to the molecule of interest is then added, the unbound Ab washed off and substrate added (see ELISA) for visualization. This assay permits specific identification of proteins in a mixture and is also often used to confirm the presence of Abs to certain infectious agents (e.g. HIV) in the serum of patients. Immunoblotting can also be used to assay for the presence of molecules in a mixture as described for the sandwich ELISA. This has now been extended for analysis of products of single cells. For example, to assay for production on a cytokine, Ab to the cytokine is coated onto the nitrocellulose 'floor' of a special culture well (see sandwich ELISA), the unbound Ab is washed off, and cells are then plated on top of this Ab.

After incubation, an enzyme linked Ab to a different determinant on the cytokine is added, followed by washing and substrate addition. Wherever a cell produced the cytokine, it will be captured by the first Ab and will then be detected by the second Ab and its conversion of substrate, forming a coloured spot on the nitrocellulose (hence the name ELISPOT assay). The nature of the cell producing the cytokine can also be determined by flow cytometry after staining the cells with a fluorescent labeled cell type specific Ab (e.g. anti-CD4 for T helper cells) and an anti-cytokine Ab labeled with a different fluorochrome.

Affinity Chromatography

Specific Purification of Ag and Ab

The specificity of Abs is not only important to the development of many research and diagnostic assays, but can, in some instances, be used to purify, or be purified by, interaction with Ag. This is because Abs do not form covalent bonds when they combine with Ag. Ab coupled to an insoluble matrix (e.g. agarose) specifically binds its Ag, removing it from a mixture of other molecules. After washing to remove all unbound molecules, the Ag can be eluted at low pH and/or at high ionic strength, which break the reversible bonds holding it to the Ab. As this can usually be performed without damaging the Ag or Ab, it is possible to obtain relatively pure Ag in one step. Similarly, Ag coupled to an insoluble matrix permits purification of Ab from media or serum.

Ab can also be purified based on its binding by proteins (e.g. protein A) isolated from some strains of *Staphylococcus aureus*. Protein A coupled to agarose binds IgG Abs which can be eluted by decreasing the pH and/or by increasing the ionic strength of the eluting buffer, again without damaging the Ab. Using similar techniques, cell subpopulations with characteristic cell surface molecules (e.g. immunoglobulin on B cells) can also be isolated (positive selection) or removed (negative selection) from a mixture of cells.

Monoclonal and Recombinant Antibodies

Monoclonal Antibodies

In 1975, Kohler and Milstein developed a procedure to create cell lines producing a predetermined, monospecific and monoclonal Ab, for which they received the Nobel Prize. This procedure has been standardized and applied on a massive scale to the preparation of Abs useful to many research and clinical efforts. The basic technology involves creation of a hybrid cell by fusion of an immortal cell (a myeloma tumor cell) with a specific predetermined Ab-producing B cell from immunized animals or people. The resulting hybridoma cell is immortal and synthesizes homogeneous, specific Ab, The utility of this Ab depends on its specificity. Monoclonal Abs (mAbs) can be made in large quantities and against virtually every Ag. Thus, mAbs have become standard research reagents and have extensive diagnostic and clinical applications.

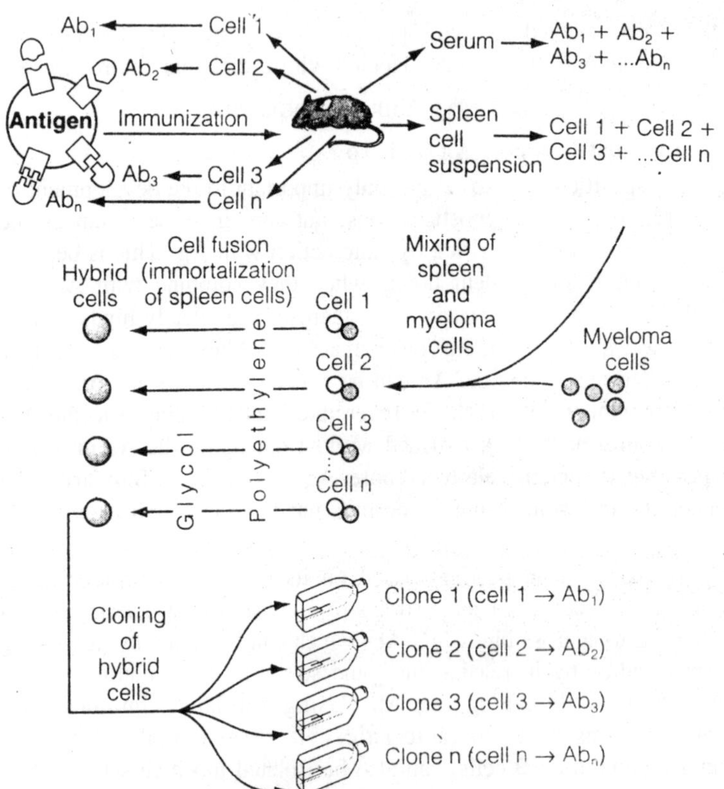

Fig. 2.8. Preparation of monoclonal abs.

Humanization and Chimerization of mAbs

The vast majority of mAbs have been developed in mice, and although useful as research and diagnostic tools, they have not been ideal therapeutic reagents at least partly because of their immunogenicity in humans. That is, murine Ab introduced into a patient will be recognized as foreign by the patient's immune system and a human anti-mouse Ab (HAMA) response will develop that compromises the utility of therapeutic Ab. This has been dealt with in two basic ways.

Humanize murine antibodies

The murine Ab can be genetically modified to be more human. In particular the constant region of the murine IgG heavy (H) and of the murine light (L) chain can be replaced at the DNA level with the constant regions of human IgG1 H and L chains to create a chimeric

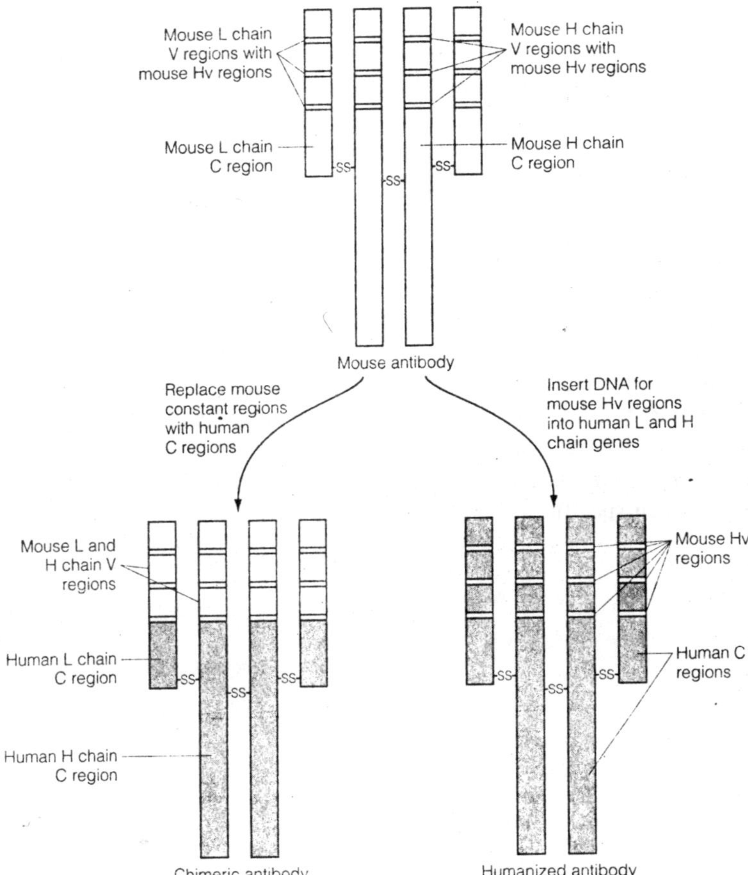

Fig. 2.9. Humanizing and chimerizing mouse monoclonal antibodies.

Ab where only the variable (V) regions are murine. This significantly decreases but does not eliminate the immunogenicity of the Ab. Another approach involves sequencing the V regions of the mouse Ab H and L chains and then inserting the DNA sequences of the hypervariable regions of these chains into human IgG H and L chain genes. The resulting Ab is 95% human with only the binding regions being murine.

Make fully human mAbs

Human Abs have been made by fusing human B cells with myeloma cells, although this has been very difficult and usually required immortalizing the B cells using Epstein–Barr virus before fusing. This approach is not ideal as a virus is used, the specificity of the mAbs produced is limited and the yield of the Abs produced is poor. More recently, a human antibody mouse has been created by replacing the genes for mouse immunoglobulins with genes for human immunoglobulins. Thus, when the mouse is immunized it makes fully human Abs against the Ag and the B cells making these Abs can be fused with myeloma cells to generate hybridomas making the human mAb.

Fv Libraries

Another way of preparing monoclonal Abs involves Fv libraries. In this approach, the H chain V regions genes of a large population of B cells are fused randomly to an equally large number of L chain V region genes to create all combinations and thus a vast number of combining sites (Fv regions). These are cloned into bacteriophage (viruses that infect bacteria) and selected for their specificity. Thus, Fvs can be expressed in a replicating bioform and used as a source from which specific mAbs can be created.

3

LYMPH GLANDS

Lymphoid cells are born, educated, and stored within specialized lymphoid organs. Lymphoid organs have a blood and lymphatic system that facilitates internal cellular movement and makes possible a ubiquitous distribution of the cells involved in immune responses. The lymphoid system includes a set of fixed cells that comprise the network of the lymphoid organ pathways of circulation and a set of mobile cells that are responsible for transporting information and instituting recognition and intervention mechanisms at a distance.

Two organs are called the "central lymphoid organs", because they give rise to the two lymphocyte populations involved in humoral and cell-mediated immunity: B cells (bursa dependent) and T cells (thymus dependent). The thymus controls the production of T cells the bursa of Fabricius, at least in birds, produces B cells. The equivalent of the bursa in mammals has not yet been discovered. The origins and development of the thymus and of the bursa of Fabricius (which occur very early in ontogenesis) are independent of antigenic stimulation. Stem cells that colonize the thymus and the bursa originate in the yolk sack and the liver, in fetal life, and in the bone marrow, in adults.

The spleen, lymph nodes, tonsils, and the gut-associated lymphoid system comprise the peripheral lymphoid system. Their development to maturation is conditioned by antigenic stimulation, as indicated by their poor development in germ-free mice (axenic mice).

B and T lymphocytes produced in the central organs migrate to peripheral organs, where they occupy preferential locations. Thus, one finds within lymphoid organs distinct functional compartments with T or B cell predominance. The analysis of these compartments may be

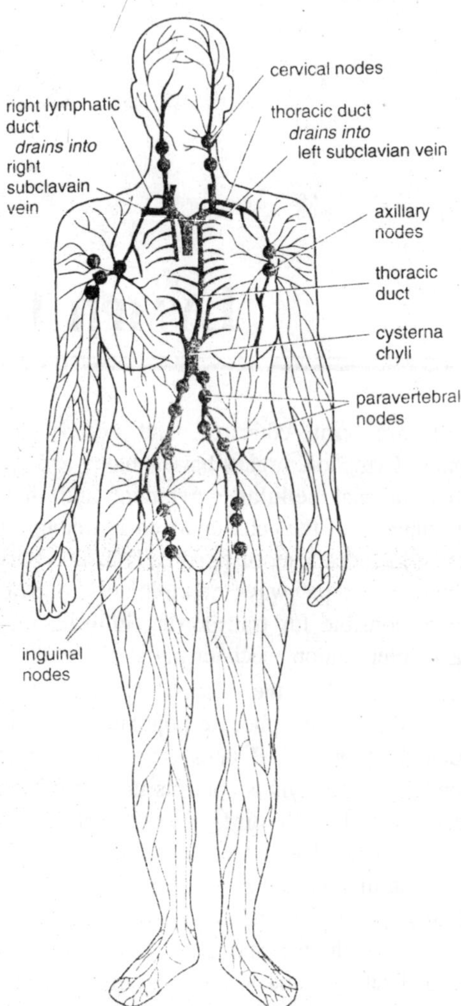

Fig. 3.1. Lymphatic vascular system.

performed by studying cellular markers specific for one cell type, in the various immunodeficiency states. Thus, the study of the lymphoid organs in athymic nude mice or neonatally thymectomized mice, by comparison with normal mice, allows definition of thymus-dependent and thymus-independent areas.

Some lymphoid cells present in blood or in peripheral lymphoid organs recirculate, that is, return to the peripheral lymphoid organs after passage through the circulatory system. To do this, they use specialized structures within postcapillary venules of lymph nodes and Peyer's patches and sinus walls of the marginal zone of the spleen. From the lymph nodes and Peyer's patches, cells travel to the circulation by means of efferent lymphatic channels and the thoracic duct. Recirculating cells reenter and remain in the while pulp of the spleen. Central lymphoid organs are excluded from recirculation. The recirculating cells are essentially long-lived T cells and prolonged external drainage of the thoracic duct (which includes 70 to 90% of T cells in rodents) or neonatal thymectomy depletes all thymus-dependent areas of peripheral lymphoid organs. However, only a small portion of competent medullary cells in the thymus recirculate between blood and lymph. Therefore, there exists a functional dichotomy of lymphoid cells, within the same morphologic structure of lymphoid organs, which depend on selective migration.

Thymus

The thymus is a voluminous, pale-white organ located in the upper anterior mediastinum at the level of origin of the major blood vessels, contiguous to the pericardium. In most species, two lobes originate from two lateral anlagen, which originally migrated to each side of the neck. The final organs are connected to each other across the midline by connective tissue. As mentioned above, the thymus in mammals is the only lymphoid organ that develops in the absence of antigenic stimulation. Axenic animals submitted to minimal antigenic stimulation show normal thymus development, whereas the peripheral lymphoid system remains atrophic. The weight and activity of the thymus are not modified by immunizations, at variance with spleen and lymph nodes.

Ontogenesis

The thymus arises from an anlage of the third and fourth endodermal pharyngeal pouches. This anlage progressively loses its ties with the pharynx, and its central lumen disappears. At this point, the thymus is a small and dense mass of epithelia cells that will next migrate downward to its definitive location in the higher thorax. In man, the thymic and parathyroid primordia are the same, which explains the symptomatology of DiGeorge's syndrome, in which the thymus and cell-mediated immunity are absent, as are the parathyroids. The thymus is the earliest lymphoid organ to appear in birds and in most mammals. The origin of thymic lymphocytes has long been controversial. It was initially believed that lymphocytes appeared locally from the epithelial network. It is now recognized that the thymic epithelial anlage becomes

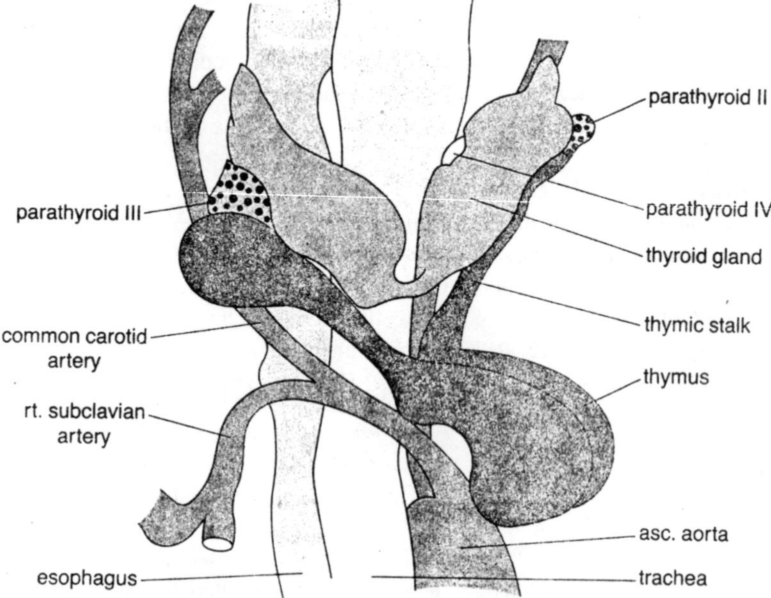

Fig. 3.2. Embryologic development of the thymus and parathyroid glands from the third and fourth pharyngeal pouches.

colonized by immigrant blood-borne stem cells that originate elsewhere. This hypothesis was suggested by Owen and Ritter, who, in 1969, showed that 10-day-old embryonic thymuses cultured in vitro never became lymphoid, while 11-day-old thymuses could become lymphoid. These results, which will be discussed later in more detail, indicate that thymic lymphopoiesis in the epithelial primordium is dependent on stem cell migration at the 11th day of gestation. The origin of these stem cells is variable during ontogenesis. They initially come from the yolk sac; when yolk sac hematopoiesis diminishes, cellular influx is produced by fetal liver; finally, the bone marrow becomes the source of stem cells in the adult. In man, the thymus completes its organogenesis during fetal life; at about 20 weeks of gestation, its aspect is that of the mature thymus. The rate of thymus growth, in relation to body weight, reaches a plateau in the third week of gestation and decreases thereafter. However, mouse thymus does not complete its development until after birth.

Structure

Each thymic lobe contains connective tissue septa that support the blood vessels and constitute lobules. Lobule size is relatively fixed

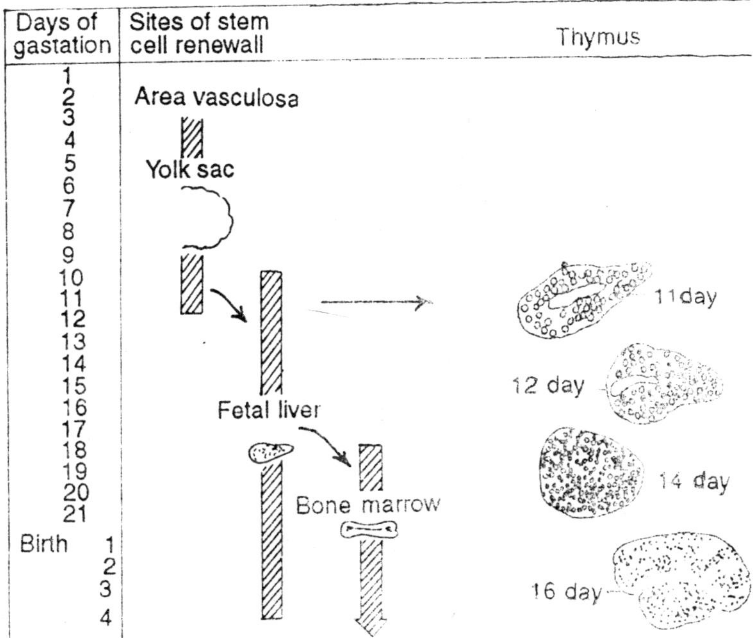

Fig. 3.3. Thymic ontogeny in the mouse.

(0.5 to 2 mm), despite highly variable thymic size in various species. These lobules are not totally separate from each other, because serial sections reveal narrow parenchymal zones that provide continuity between lobules. Each lobule features a periphery, very rich in lymphocytes, a cortex, and a medulla, which is less abundant in lymphocytes and shows weaker staining.

The thymus is nourished by vessels that arise from the internal thoracic arteries. The arteries branch at interlobular septa and penetrate lobules at the level of the corticomedullar junction. The cortex is vascularized by arterial capillaries with anastomotic arcades; the arterioles join postcapillary venules at the corticomedullary junction. The medulla contains arterioles and venules. Some lymphatics may be seen within the connective tissue, but the parenchyma itself is not penetrated by lymphatic vessels. The thymic parenchyma consists of a network of epithelial cells, which delineate spaces, without interstitial tissue, in which lymphocytes accumulate, especially in the cortex.

Epithelial network of the thymus

The thymic network is comprised of epithelial cells of endodermal origin. In the cortex, these cells are elongated and have cytoplasmic extensions between lymphocytes. Epithelial cells are larger and more numerous in the medulla. The study of epithelial cells by electron microscopy shows linkage by desmosomes; some of the tonofilaments in the cytoplasm are inserted into the desmosomes. These epithelial cells are contiguous to small numbers of macro-phage-type reticular cells of mesenchymal origin. Epithelial cells possess secretory granules in their cytoplasm, which suggests that they may play an important role in the secretion of lymphocyte maturation-inducing thymic factors. This hypothesis has been corroborated by functional studies that show that the thymus or purely epithelial thymoma grafts restore the immunologic competence of neonatally thymectomized mice and by the ability of in vitro cultured epithelial cells to induce the appearance of T cell markers in thymectomized mouse lymphocytes.

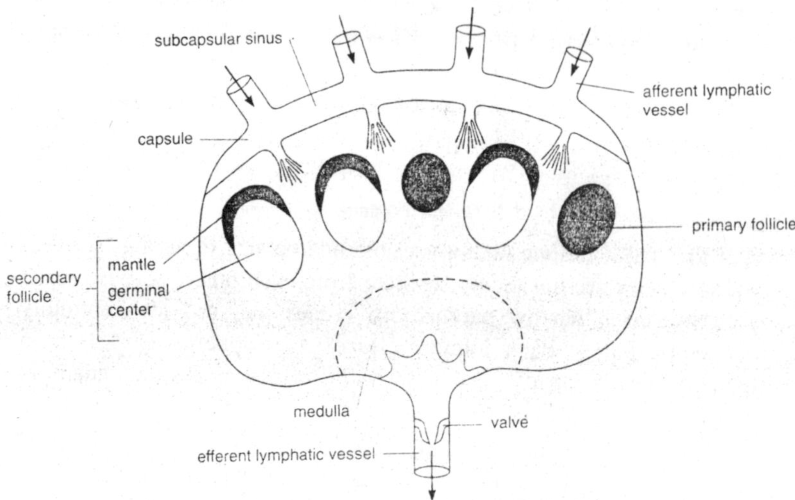

Fig. 3.4. Idealized structure of a lymph node. Lymph enters via afferent vessels, passes through the cortex (shaded) and medulla, and exits via a single efferent vessel.

In the medulla, epithelial cells are grouped in round structures called Hassal's corpuscles. In some species, for example, the mouse, corpuscles are, instead, simple aggregates of epithelial cells. Hassal's corpuscles consist of cells piled and coiled on top of one another. Their size is variable, and they may be cystic, with a central eosinophilic substance plus pyknotic cells. Electron microscopy confirms the epithelial nature of the cells that form these corpuscles. Here, too, they are linked together by desmosomes and show cytoplasmic tonofilaments. In the middle of the corpuscle may be seen nuclear pyknosis, a decreased number of intracytoplasmic organelles, and large amounts of keratohyalin, which are characteristics similar to those of skin epithelial cells. The peripheral cells contain microvilli directed toward the corpuscle's center, sometimes thick enough to form a brushlike border. The functional significance of Hassal's corpuscles remains obscure. The presence of active cell with numerous microvilli and the mode of vascularization, which resembles that of endocrine glands, suggests that Hassal's corpuscles may be active organelles rather than degenerative structures. They could conceivably playa role in transferring the humoral products of the thymus to the circulation.

Thymic cortex

The thymic cortex is densely infiltrated by lymphocytes of various sizes. Large lymphocytes are less numerous and are mostly located in the subcapsular cortex. Moreover, in this zone, the stem cells penetrate and divide, explaining the high incidence of mitoses. It is interesting to note that small lymphocytes of the superficial cortex have already acquired T cell-specific membrane antigens (θ and TL antigens in the mouse). The deep cortex stores small lymphocytes. One may also observe cellular destruction in the deep cortex: a large number of cortex produced lymphocytes die locally.

Medulla

Very few lymphocytes are present in the medulla, therefore, its epithelial framework is easily visible alongside Hassal's corpuscles. Medullary lymphocytes are the most nature thymic lymphocytes. They include lymphocytes that migrate from cortex to medulla and then leave the thymus (although direct migration of cortical cells into the blood stream has recently been suggested). In mice, medullary lymphocytes differ from cortical lymphocytes by several criteria. They do not bear the TL antigen, they respond to phytohemagglutinin (PHA) and concanavalin A (con A), and they are capable of inducing the graft-versus-host reaction. They are steroid resistant, which permits their selective isolation after destruction of cortical thymocytes by hydrocortisone treatment.

Lymphocytes leave the thymus by passing between cells of the medullary venule wall. Raviola and Karnovsky have shown that tracers injected into the systemic circulation may be found in the thymic medullary parenchyma. Conversely, cortical arterioles are impermeable to tracers, cortical vessels contribute to the building of a barrier between blood and cortical cells. These data, along with knowledge of the vascular system of thymic lobules, suggest that only medullary cells come into contact with antigens. The thymus, unlike peripheral lymphoid organs, does not normally contain lymphoid follicles or germinal centers; these structures are found only in pathologic states, such as lupus erythematosus and myasthenia gravis. In these disease states, they are exclusively present in the medulla, where antigenic stimulation may occur.

Involution

The thymus involutes with age. This involution begins at a precocious age; for example, in the mouse, the thymus weight begins to decrease at 6 weeks of age, first rapidly, then more slowly. Thymus atrophy initially occurs in the cortex, which become progressively thinner and the parenchyma is then infiltrated by adipose tissue. In man, thymus involution begins at puberty, as shown by a reduction in the corticomedullary ratio. However, the thymus never completely disappears. Even in old age, one finds, on studying numerous histologic sections, islets of thymic parenchyma that contain a few lymphocytes within the adipose tissue that has invaded the thymus.

Bursa of Fabricius

The bursa of Fabricius is a lymphoid organ located at the terminal portion of the cloaca in birds. Like the thymus, it is lymphoepithelial. It is also a primary lymphoid organ whose development is independent of exogenous antigenic stimulation.

Ontogenesis

The bursa of Fabricius is the second lymphoid organ after the thymus, to appear in birds. On the fifth day of incubation, it appears as an invagination of the posterior wall of the cloaca. Around the 10th to the 12th days, the epithelial cells that line the "evagination" proliferate into epithelial buds, which arise within the underlying connective tissue. On the 12th to 13th days of incubation, large basophilic cells, the lymphoid stem cells, appear in the epithelial primordium. The origin of bursal lymphocytes is as controversial as the origin of thymocytes. Moore and Owen showed, in parabiotic animals with chromosomal markers, that bursal lymphocytes were not endogenous but represented migrating cells brought to the bursa by the bloodstream, probably originating in the yolk sac.

The bursa of Fabricius is the locus of differentiation of B lymphocytes that produce antibodies. Lymphocyte maturation begins early in ontogenesis. Bursectomy on the 17th day of incubation induces agammaglobulinemia, with absence of germinal centers and plasma cells in peripheral lymphoid organs.

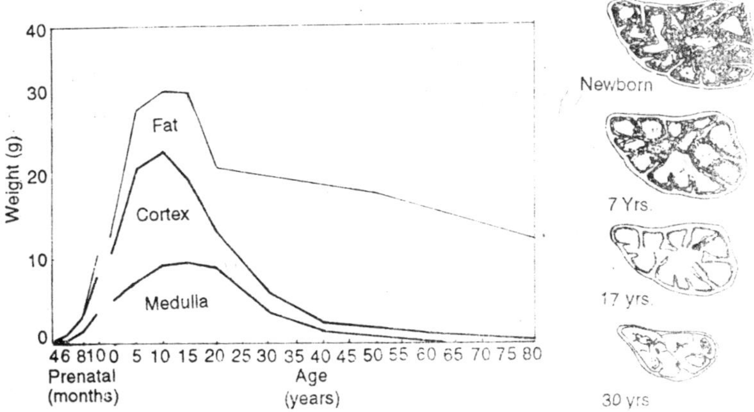

Fig. 3.5. Thymic involution with age in man.

Structure

The bursa of Fabricius is an asymmetric medial organ that opens into the posterior wall of the terminal intestine. It is shaped like a sack and contains a star-shaped lumen that is continuous with the cloacal cavity. Its maximal diameter is 3 cm in the chicken. The mucosa, the musculosa, and the serosa of the bursa are in continuity with the corresponding tissues of the intestine. The epithelial surface, like that of the intestine, consists of cylindrical cells, but the bursa has no mucous cells. Lymphoepithelial nodules are present in the lamina propria directly under the epithelium. These nodules contain a light medulla and a dark cortex. The medulla contains epithelial cells that form a continuous area in the periphery, which projects into the epithelial coating. The center of the medulla is less structured, it

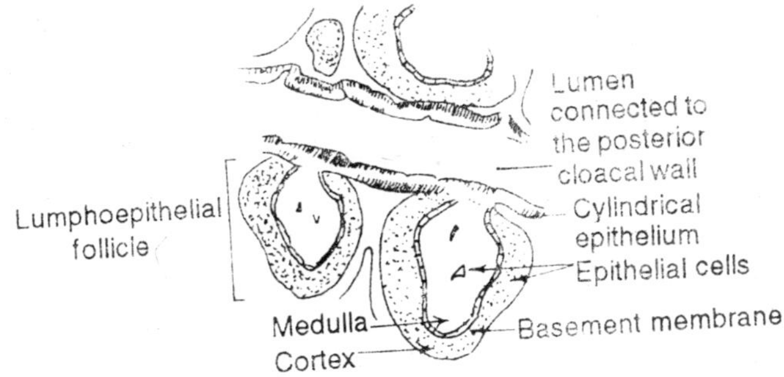

Fig. 3.6. Structure of the bursa of Fabricius in the chicken.

contains, in addition, epithelial cells and various other types of cells, including large lymphocytes, plasma cells, reticular cells, macrophages, and granulocytes. Unlike the thymus, in which no precise separation exists between cortex and medulla, there is present in the bursa of Fabricius a 100- to 140-nm basement membrane, which separates the medullary epithelium from the cortex. On micrographs, it looks like a central homogeneous zone 15 mm in diameter and is bordered on both sides by an amorphous glycoprotein layer. This membrane is not completely impermeable, because it permits exchange between the two zones of the lymphoepithelial nodule. The cortex consists mostly of small lymphocytes and plasma cells.

Involution

Like the thymus, the bursa involutes rapidly with age. At 4 months, a time that corresponds to puberty in the chicken, the bursa begins to atrophy and practically disappears by the end of the first year. Bursal regression is related to increases in testosterone or estrogen levels at puberty..Premature regression occurs after administration of testosterone. If administered in ovo between 5 and 8 days of incubation, testosterone may eliminate bursal development.

Bursal equivalent in mammals

A bursal equivalent has been sought in mammals for a long time. The equivalent tissue might be responsible for B cells differentiation. B cell precursors are found in the mammalian fetal liver and in adult bone marrow. Nothing, however, is known of the microenvironment necessary for B cell maturation in mammals. Some investigators suggested that the gut-associated lymphoid system may be central B cell maturation organ. However, there is considerable evidence to the contrary, and no real bursal equivalent is recognized in mammals at this time.

Bone Marrow

In the strict sense, this structure is not just a lymphoid organ. It has major importance, however, because the bone marrow produces precursor cells of various lymphocyte populations and macrophages in the adult. A single bone marrow cell injection completely restores the lymphoid system of irradiated (but not thymectomized) rats or mice. Bone marrow has a ubiquitous distribution and fills the free spaces inside bones.

Ontogenesis

Bone marrow is initially composed of mesenchymal primitive elements. The precise onset of hematopoiesis varies among bones and species. In man, active zones of hematopoiesis are observed in the clavicle of l0-week-old (43 mm) embryos, whereas hematopoiesis does not occur in the femur until the 14th week (75 mm). Bone marrow reaches full hematopoietic activity only at mid-gestation, at which time liver hematopoiesis begins to regress.

Structure

Bone marrow represents a complex vascular system within the hematopoietic tissues.

Vascular system

This system consists of an afferent artery, a capillary network, and venous sinuses. The afferent artery perforates the cortical bone and then divides into an arterial capillary network that communicates directly with venous sinuses. These venous sinuses are highly developed: they are bordered by endothelial cells, which may be phagocytic and have a glycoprotein basement membrane. The basement membrane is coated by a layer of adventitial cells that possess extensions that enter the hematopoietic compartment. Sinus walls are discontinuous, thus allowing cellular exchange between blood and tissue. Venous sinuses open into longitudinal central veins.

Hematopoietic tissue

Hematopoietic tissue includes all circulating blood cell lines and their precursors: erythroblasts (grouped in islets around one or two reticular cells), granulocytic cells, monocytes, megakaryocytes, and lymphoid cells including a few plasma cells. Lymphocytes comprise as much as 20% of bone marrow cells, especially in rodents the number of lymphoid cell precursors is not established, because they are difficult to identify. The proportion of various cells varies, however, according to peripheral needs.

Lymph Nodes and Lymphatic System

Lymph nodes are round, or reniform, lymphoid organs, that consist of a parenchyma infiltrated by lymphocytes and surrounded by a capsule. Nodes may be single but more often are grouped along the pathway of a lymphatic vessel.

Ontogenesis

Embryonic development of lymph nodes is closely linked to the lymphatic system. The lymph node parenchyma develops from the mesenchyma, which surrounds the lymphatic plexus around primitive lymphatic pouches and which is very early infiltrated with stem cells.

In man, organized lymph nodes may be recognized in 50 mm embryos (11 to 12 weeks) in axillary and ilia areas, stem cell aggregates may be seen much earlier (in the 30 mm embryo).

Lymphatic Vascular System and Lymph Node Lymphatic Circulation

The blood capillary network is surrounding by a lymphatic network. Lymphatic capillary vessels ending blindly carry lymph from extravascular spaces to the blood. Lymphatic capillaries become channels of a larger diameter that end in large collector vessels, the thoracic duct and the right lymphatic cord, which, in him, enter large blood vessels in the neck. The capillary wall is very thin, representing only one layer of endothelial cells, the edges of which may overlap. The absence of a basement membrane permits exchange with the interstitial medium. The larger lymphatic vessels have a thicker wall, which contains collagenous and elastic fibers and, sometimes, smooth muscle fibers, and an internal elastic membrane. However, the distribution of different layers is not as well structured as in blood vessels. These lymphatics have valves with the free edge pointing toward the direction of flow, which prevent reflux. Each series of valves is associated with widening of the vascular diameter, thus giving an irregular appearance to the lymphatics. The absence of an organized vessel wall explains why lymph flow is dependent on mobilization by adjoining structures, particularly muscular contractions. The lymph nodes that connect to these vessels receive their lymph by means of vessels that penetrate the capsule. The capsule and the lymph node parenchyma are separated by a peripheral sinus sending out intermediary sinuses along fibrous trabeculae. These intermediary sinuses enter the cortical parenchyma and then branch into medullary sinuses separated by medullary cords. The sinuses enter the efferent lymphatic pathway, which leaves the lymph node at the thelum. The sinus wall is bordered by endothelial cells that rest on a reticular network that is continuous with the parenchyma. Fibers bridge the sinuses and hold star-shaped cells and macrophages. This sinus architecture represents a remarkably efficient filtration and exchange system between lymph node parenchyma and the lymph.

Lymph Node Blood Supply

An artery enters the lymph node at the hilum. Its branches follow the fibrous trabeculae, penetrate the medullary cords, and reach the cortex, where they fan out into a terminal capillary at the level of the lymphoid follicle. Postcapillary venules extend from the capillaries into the deep cortex. Their wall consists of endothelial turgescent cuboidal cells without a muscular layer. Circulating lymphocytes leave the capillaries, pass between the large cells of postcapillary venules, and also may pass through endothelial cells. Veins exit from the lymph node at the hilum. Similar vessels are found in other lymphoid organs, such as Peyer's patches, tonsils and appendix.

Lymph Node Parenchyma

The lymph node parenchyma consists of a network of reticular fibers and reticular cells within which motile cells, including lymphocytes, plasma cells, and macrophages, are trapped. The lymph node contains a cortical zone and a lighter and less cellular medullary zone that contains numerous sinuses. Two zones can be distinguishes within the cortex: the external cortex, also called the subcapsular cortex, where lymphoid follicles with germinal centers develop, and the deep cortex, also known as the paracortical or diffuse cortex.

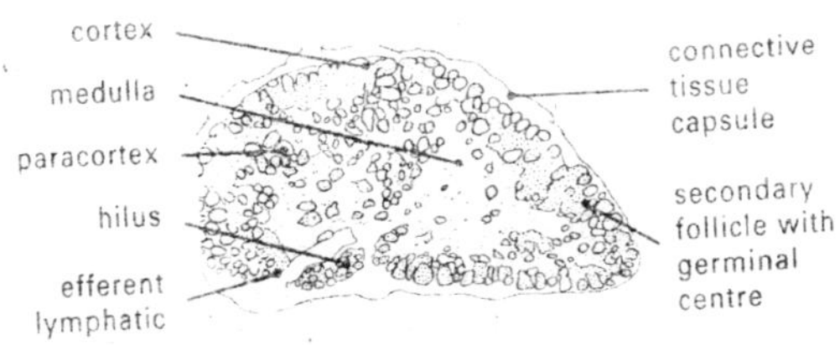

Fig. 3.7. Lymph node section.

External cortex

Lymphoid follicles and their germinal centers are located in the external cortex. The ollicles without germinal centers are "primary follicles", whereas those with a germinal center are "secondary follicles". Secondary follicles, when correctly oriented, have a polarity directed toward the capsule. Within the germinal center, there are two different zones. The portion close to the capsule is less cellular. The pole opposite from the capsule is denser (darker). In the dark zone, the tightly packed lymphoid cells show numerous mitoses. Various phases of evolution are apparent including lymphoblasts, large and medium lymphocytes, and young cells of the plasma cell series. Macrophages contain phagocytosed cellular debris. This is the fertile zone of the germinal center. The light zone of the germinal center does not contain cells undergoing mitosis and is progressively replaced by small lymphocytes. The young cells have basophilic cytoplasm in the dark zone. All cells rest on a network of anastomosed dendritic cells. In the periphery reticular fibers are distributed concentrically

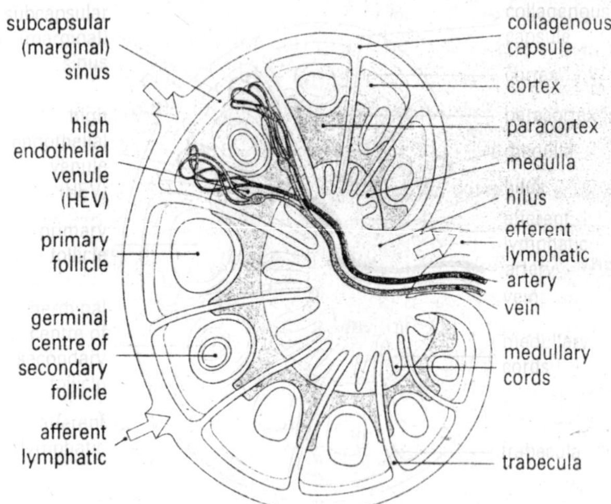

Fig. 3.8. Schematic structure of the lymph node.

and disappear into the germinal center. The germinal center is ringed by lymphocytes, which are particularly numerous on the top of lighter pole of the center, causing a "cap" appearance, which diminishes closer to the fertile zone.

Follicular cells belong to the B cell lineage. Follicles are not depleted by thymectomy or by chronic cannulation of the thoracic duct (during which procedures recirculating long-liver lymphocytes are depleted). They show membrane immunoglobulins, by immunofluorescent staining.

In addition to B cells, germinal centers contain rare T cells. The presence of these T cells is critical for the formation of the germinal center. In artificially T cell-depleted mice (thymectomized, irradiated, and bone marrow reconstituted), and in natural T cell deprived nude mice, germinal centers are infrequent or absent.

Deep cortex of the lymph node

The deep cortex contains postcapillary venules through which lymphocytes travel. This part of the cortex is mostly composed of T lymphocytes. This zone is depleted after neonatal thymectomy or by thoracic duct cannulation. In nude mice, the deep cortex is present but does not contain lymphoid cells.

Medullary parenchyma

The medullary parenchyma contains cordlike structures with numerous ramified sinuses and is of mixed cellularity. Macrophages, plasma cells (which migrate into the efferent lymph immunization), and also T lymphocytes are present in this zone. We shall see later that the paracortical zone is mostly populated by T lymphocytes, whereas lymphoid follicles are essentially thymus independent.

Lymph Node Modifications During Immunization

The relative volume of the above areas within the lymph nodes (including cortex, deep cortex and medulla) varies depending on quiescence or immunologic activity. Immunization alters the relation and absolute size of B and T cell-dependent areas, depending on the type of immune response elicited.

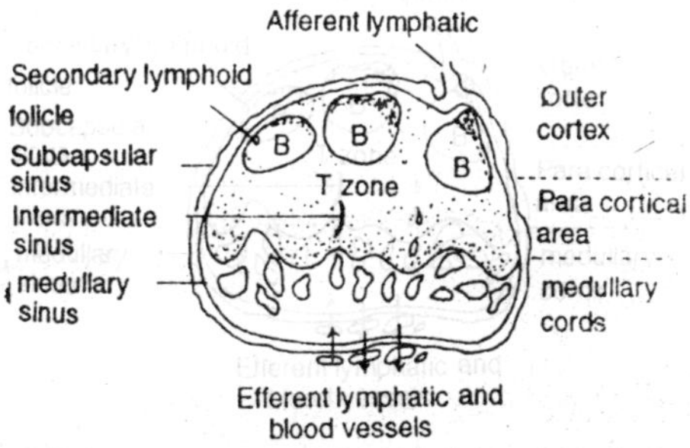

Fig. 3.9. Structure of a peripheral lymph node in the mouse.

Cell-mediated immunity

The lymph node that drains the skin area of guinea pig and mouse ear after local application of a cell-mediated immunity stimulant, such as oxazolone, undergoes characteristic modifications. Islets of pyroninophilic blast cells are seen in the deep cortex at 24 hr. These cells then proliferate up to Day 4, causing an increase in the volume of the deep cortex, while the medulla is compressed. Postcapillary venules show significant changes during the 2 days after immunization. Endothelial cells increase in size, and many lymphocytes cross the

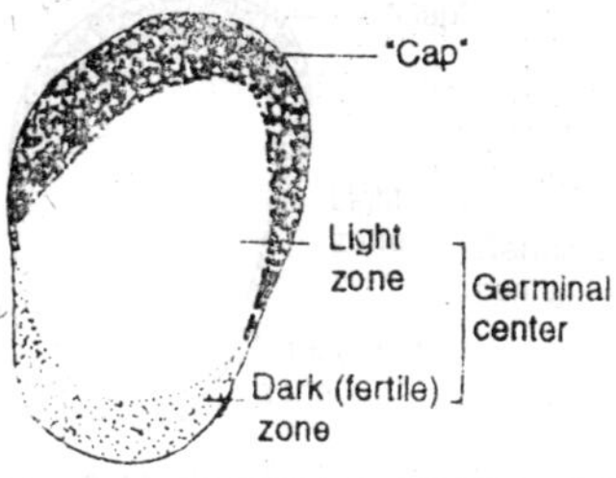

Fig. 3.10. Structure of germinal centres.

capillary wall. Activity diminishes when blast proliferation starts in the deep cortex. Blast cells are still present on Day 4 but have completely disappeared by the seventh day, after giving rise to a new population of small lymphocytes. The cortical zone and its follicular crown are very little modified by immunization.

In neonatally thymectomized animals, lymph nodes remain small, and cellular proliferation is not seen in the deep cortex (which is deficient in lymphocytes). Cell-mediated reactions are not always so selective: the changes described above are not unique and are followed, at the beginning of the second week, by a reaction of the thymus-independent zones, with germinal center formation and appearance of immature and mature plasma cell in the medulla.

Humoral responses

Immunization by a thymus-independent antigen that gives rise to antibody production, such as polysaccharide, provokes changes mostly in the thymus-independent zones of lymph nodes. The injected antigen is very rapidly found in the medulla. It is then taken up by the lymphoid nodules of the superficial cortex is less than 1 hr for a primary immunization and in less than 10 min for a secondary immunization. The antigen is found in the marginal zone of the follicle, linked to: reticular cells whose ramified projections penetrate between lymphocytes. The antigen is then attracted toward the follicle's center, where it induces proliferation, starting in the fertile zone of the germinal center. The newly formed cells, which progressively assume the appearance of immature plasma cells, migrate toward the apical light pole of the germinal center between Days 5 and 6. Lastly, an influx of mature plasma cells is observed in the medulla. Only few modifications occur in the deep cortex throughout immunization.

Spleen

The spleen is a voluminous filter placed on blood vessels. It retains particles and cellular debris carried by the circulation, collects antigens, and is the locus of B or T cell-mediated reactions. In some vertebrates, it also plays a role in the formation of granulocytes, erythrocytes and platelets.

Ontogenesis

The spleen appears initially as a mesenchymal thickening of the left posterior edge of the stomach in 8- to 10-mm human embryo (fifth week). This mesenchymal tissue includes star-shaped cells that will form the reticular web of the spleen in early mammals, it consists of a capillary network connected to afferent arteries and efferent veins. Irregular spaces, which subsequently become sinuses, appear and make contact with the vascular system. Later on, at variable periods before and after birth perivascular areas are infiltrated with circulating lymphoid cells; these cells represent the initial primordium of the white pulp, whose definitive organization is only achieved after birth.

Vascularization and organization

The splenic capsule sends fibrous trabeculae that penetrate and partition the organ. The capsule and connective tissue walls contain elastic and smooth muscle fibers that, in certain species, play an important role in altering the volume of the organ. Branches of the splenic artery enter the hilum and penetrate the splenic parenchyma along the connective tissue walls. Splenic arteries divide and become surrounded by a lymphoid sheath.

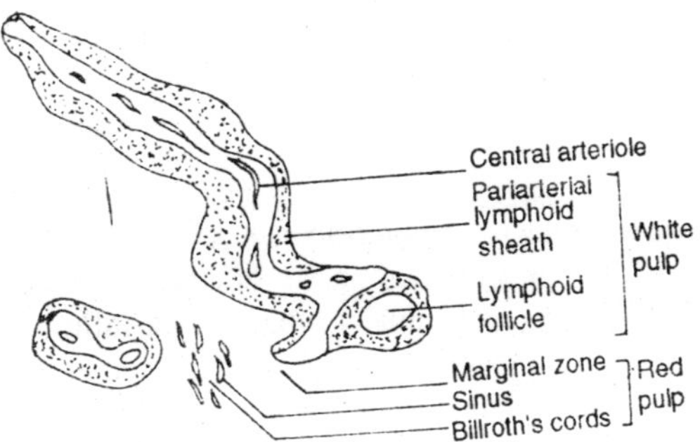

Fig. 3.11. Human spleen.

The periarterial lymphoid tissue constitutes the white pulp of the spleen and is visible macroscopically as grey zones, contrasting with the red pulp. On sectioning, the white pulp shows classic malpighian corpuscles. Splenic arteries branch to form "penicillaries" arterioles that lose their lymphoid sheath and divide into two or three capillaries that enter the venous sinuses of the red pulp. These venous sinuses are vascular spaces bordered by an edothelium that rests on a fenestrated basement membrane through which important cellular exchanges occur. The sinus wall is in continuity with the reticular tissue of splenic cords (Billroth's cords). Venous sinuses, vessels, and Billroth's cords constitute splenic red pulp. Venous sinuses open into pulp veins that penetrate and fibrous trabeculae and are drained by hilar veins that are tributaries of the splenic vein.

Lymphatic vascularization is little developed and poorly understood. In some mammals, a lymphatic efferent pathway runs along the central arteriole of the white pulp. This pathway may be the immigration route of lymphocytes in these species.

Red pulp

We have seen that the red pulp contains vessels, venous sinuses, and Billroth's cords. These cords consist of a reticular fiber network that supports the star-shaped cells that cling to the sinus walls. Some of these star-shaped cells are fixed macrophages, which play an important role in eliminating sick cells and particles in the blood. The cellular network thus formed between sinuses contains mobile cells, including, in variable proportions, depending on the functional state of the spleen, numerous macrophages with inner debris, plasma cells, lymphocytes, erythrocytes and granulocytes. The spleen is a hematopoietic organ in the embryo and even in the adult for certain mammalian species. In the latter case, it contains erythroblastic islets, young cells of the myeloid lineage, and megakaryocytes. The red pulp also plays a role in the capture of antigens that are initially bound to macrophages in the marginal zone and the rest of the red pulp.

Marginal zone

The marginal zone is an intermediate area between the red and white pulps and contains numerous sinuses oriented concentrically around the periarterial zone. Its proximity to the arterial influx explains why cells and/or antigens injected into the systemic circulation are very rapidly accumulated. It contains more lymphocytes and plasma cells than does the red pulp. It contains the zone of exchange with the white pulp. The walls of the marginal sinuses in the marginal zone have the same function as postcapillary venules in lymph nodes. Thus, lymphocytes labeled with tritiated thymidine and injected into the systemic circulation are found in the marginal zone within 15 to 30 min and are present in the peripheral zone of the white pulp within 24 hr. Red pulp and marginal zone contain both B and T lymphocytes (mixed zones).

White pulp

The white pulp is organized around central arteries. Lymphocytes that surround these arteries are enclosed in a reticular network that constitutes several peripheral layers and thereby isolate this peripheral zone from the marginal zone. The reticular network contains several "windows" that permit cellular and particle exchanges between the periarterial and marginal zones. The periarterial lymphocyte layer is composed essentially of small and medium lymphocytes. The periarterial reticular layer includes lymphoid follicles with their germinal centers distributed excentrically to the central artery. Ramification of this artery ensures the vascularization of the follicles, whose structure is similar to that described above for lymph node follicles. Splenic follicles are oriented in such a way that the light zone of the germinal center and the lymphoid crown are turned toward the red pulp; the fertile zone is oriented toward the central artery.

As in lymph nodes, the white pulp of the spleen contains thymus-dependent and independent zones. The periarterial thymus-dependent areas are depleted of lymphocytes in neonatally thymectomized mice and in mice thymectomized, irradiated, and bone marrow reconstituted. Conversely, the external lymphoid sheath and follicles are thymus independent.

Modifications of the spleen during immunization

Antibody production is associated with numerous modifications of splenic red and white pulps. Antigen binding occurs first, almost immediately in the red pulp and in the marginal zone at the level of the dendritic projections of reticular cells. The antigen is very rapidly concentrated in the marginal zone and after 2 hr is found in the white pulp at the periphery of lymphoid follicles. Antigen movements inside the germinal center occur as described above. Within 24 hr, the marginal zone is cleared of antigen, which is then concentrated in the germinal centers. The thymus-dependent periarteriolar zone of the white pulp, however, remains free of antigen. The cellular events that accompany or follow antigen binding in the spleen involve various organ compartments. The appearance of blast cells in the thymus-dependent periarteriolar zone, cellular proliferation and the appearance of young plasma cells in the germinal centers and of intermediate and mature plasma cells in the germinal centers in the marginal zone, and in the red pulp. Cellular events in the white pulp precede those that occur in the red pulp and in the marginal zone. The intensity and rapidity of modifications within the thymus-dependent areas, lymphoid follicles, and their germinal centers vary according to the route of immunization and the primary or secondary nature of the response.

Tonsils

Tonsils are lymphoid organs circumferentially placed around the pharynx. In man, the palatine tonsils are located between the columns of the soft palate, lingual tonsils are located at the back of the ventral tongue, tubal tonsils are present adjacent to eustachian tubes, and the pharyngeal tonsils are located on the posterior wall of the pharynx. Taken together, these tonsils constitute Waldeyer's circle. Tonsils are lymphoid aggregates that contain lymphoid follicles with germinal centers identical to those of the lymph nodes. Their light zone is oriented toward the epithelium. Tonsillar tissue is immediately adjacent to malphighian-type epithelium, whose crypts penetrate deeply into the lymphoid tissue. Immunofluorescence studies show that IgA and IgC cells are preferentially located in the mucosa, whereas IgM cells are found in the germinal centers.

Gut-associated Lymphoid System

Anatomy

There are three general localizations of lymphoid cells in the gut: the digestive mucosa itself. Lymphoid organs regularly distributed along the small intestine, ending in the vicinity of the appendix (when grouped together within the small intestine, these organs constitute Peyer's patches at the antemesenteric edge of the intestine. Rodents have nine to 11 Peyer's patches, but in man their number varies according to age, reaching a maximum of approximately 300 at 12 years) mesenteric lymph nodes that are distributed throughout the mesentery.

Digestive mucosa

The intestinal mucosa includes connective tissue, lamina propria that separates Lieberkuhn's glands, which underlie the small bowel villa, and an external epithelium. The lamina propria in adults contains numerous plasma cells and T lymphocytes. Only IgA-type plasma cells are found in the mouse and the rat. In man, plasma cells are mostly IgA, but there are also some IgM and IgC plasma cells. Lymphocytes are distributed among epithelial cells, usually adjacent to the basement membrane. These lymphocytes do not contain surface immunoglobulins detectable by immunofluorescence, and this fact suggests a thymic origin. This suggestion has been confirmed by evidence, in mice, that they bind rabbit anti-T heterospecific antisera.

Peyer's Patches

Peyer's patches were once considered the bursal equivalent of mammals. They are, in fact, a peripheral lymphoid organ that contains both B and T cells. They make their appearance late in fetal life (man) or only after birth (mouse). In the adult, they consist of lymphoid follicles with very large germinal centers. Analogous to lymph nodes, their germinal centers contain a few T lymphocytes. However, the great majority are rapidly dividing B cells (it is possible to label

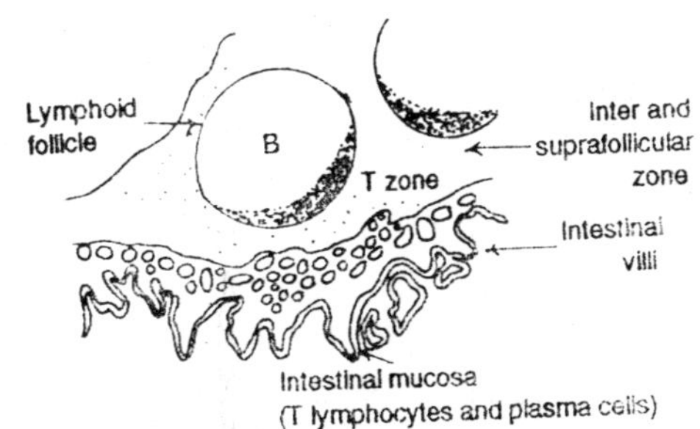

Fig. 3.12. Structure of a Peyer's patch in the normal mouse.

them in 1 hr in vitro by adding tritiated thymidine). They are immature cells without plasma cell differentiation. Half of them bear α chains on their surface. Adjacent to the follicles of Peyer's patches, the flattened mucous membrane contains both T and B lymphocytes interspersed among epithelial cells. Follicles are separated by mucosal areas coated with T cell-infiltrated villa. These areas are extremely hypoplastic in nude mice and in neonatally thymectomized mice. A few IgA, IgM, and IgG plasma cells are located at the periphery of the patches.

Mesenteric lymph nodes

Mesenteric lymph nodes have the same structure as other lymph nodes, including a superficial thymus-independent cortex and a deep cortex that contains T cells and a plasma cell-rich medulla. Unlike peripheral (and axillary or popliteal) lymph nodes, mesenteric lymph

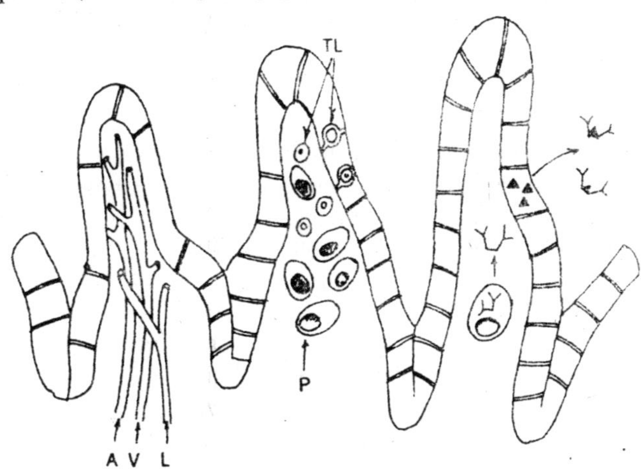

Fig. 3.13. Intestinal villi. A—Artery; V—vein; L—lymphatic; P—plasma cell; TL—T lymphocyte.

nodes experience constant antigenic stimulation, as indicated by the number and size of germinal centers found in the superficial cortex and the medulla. Half of these rapidly dividing B cells of mesenteric lymph nodes contain α chains (analogous to Peyer's patches), medullary plasma cells are mostly of the IgA type.

Lymphatic vessels connect the various gut-associated lymphoid organs. The lymph that drains the intestinal wall is collected in centrovilli chylifers that join to form a subserous network, particularly dense at the level of Peyer's patches. This network then gives rise to lymphatic trunks that constitute the efferent vessel of mesenteric lymph nodes. Nearly all gut-produced lymph is thus carried to mesenteric lymph nodes. Efferent vessels of mesenteric nodes form a unique network that opens directly into the thoracic duct. Only the lymph from the lower colon connects to lateroaortic and inguinal nodes. Thus, intestinal lymph, after a stop in mesenteric nodes, enters the thoracic duct and proceeds to the venous and arterial circulations.

Cycle of IgA Bearing Cells

IgA plasma cells in the lamina propria are mature short-lived cells (4- to 5- day half-life in the mouse) that are also secretory. Precursors of these plasma cells, recognized by the presence of immunoglobulins are found in germinal centres of Peyer's patches and in mesenteric lymph nodes. It is possible to repopulate IgA plasma cells in irradiated rabbits by injecting cells from Peyer's patches. Proliferation of precursor cells depends on at least two factors: local antigenic stimulation and B- and T- cell collaboration.

Precursor proliferation

Local antigenic stimulation. Germinal centres in Peyer's patches and IgA plasma cells (intestine) develop after the appearance of normal intestinal flora (after weaning in the mouse) and are absent in axenic mice. Moreover, only gut immunization induces are appearances of IgA plasma cells in the mouse. In man, antipolio vaccination provokes antibody formation only in the mucous secretion when a living attenuated virus is administered orally.

Need for B and T-cell collaboration. There are few or no plasma cells in the intestinal mucous membrane of nude mice (there are, however, a precursors). Thymus grafting induces both an increase in T cells in Peyer's patches and the appearance of germinal centers.

IgA-cell migration

The IgA cell population of the digestive mucous represents an active cell migration. Transfer experiments and analysis of the cellular content of various lymphoid organs has demonstrated the existence of a true IgA-cell cycle. The precursors leave Peyer's patches by mean of lymphatic vessels, undergo a progressive maturation, and transform into IgA-containing mature cells in mesenteric nodes which then enter the thoracic duct and finally preferentially localize in the intestine and, to a lesser degree, the spleen.

The reasons why mature IgA cells are attracted to the intestine are only hypothetic. An attraction to the antigens of intestinal flora is not likely. It is possible that there is a role of intestinal receptors for the IgA secretory piece, this possibility has not, however, been demonstrated.

T cells appear in the intestinal mucous membrane rather late, after plasma cells. The origin and role of these cells are also unknown.

Conclusions: Lymphoid System Organization and the Homing Phenomenon

The comparative morphologic study of the lymphoid system in normal and thymus-deprived animals and the study of morphologic changes associated with cell-mediated immunity and humoral reactions demonstrate that the peripheral organs contain thymus-dependent and independent zones. Thymus-dependent zones include the deep cortex of the lymph node (paracortical area), the periarterial lymphoid sheath of the splenic white pulp, and inter and suprafollicular zones within Peyer's patches. Thymus-independent zones include the follicles of the external cortex of lymph nodes and Peyer's patches, and the follicles of the peripheral zone of the splenic white pulp. Splenic red pulp, particularly the marginal zone and the medulla of lymph nodes, contain predominantly B cells, but they are also areas where B and T cells circulate, these regions are therefore considered to be mixed territories.

The phenomenon of selective migration of B and T lymphocytes to distinct zones of peripheral lymphoid organs is as yet unexplained. This remarkable ability of lymphocytes to recognize and colonize privileged locations was termed "homing", or "ecotaxis", by DeSousa; it has been studied by following the localization in peripheral organs of cells of various origins, after radiolabeling and injection into syngeneic hosts. Thoracic duct and thymic lymphocytes migrate preferentially to thymus-dependent areas. Spleen cells are of heterogeneous composition (B and T cells) and distribute in several B and T peripheral compartments. Lymph node cells essentially migrate to thymus-dependent areas but a certain proportion of cells migrates in follicles of the external cortex. Bone marrow cells migrate to thymus-independent areas, the red pulp of the spleen and lymph node medullar cords. Some marrow cells, poorly differentiated, return to the bone marrow, as myeloid cells.

Several hypotheses have been proposed to explain "homing". Structural differences in the reticular network of the various' compartments have been suggested. The looser web seen in thymus-dependent zones might select out a particularly mobile cell type. Chemotaxis, which plays a role in the mobilization of certain cells (like polymorphs), has not been demonstrated for lymphocytes. The property of cellular adherence could also playa role, because thymus-dependent cells, which are nonadherent, show a different mobility than that of thymus-independent cells (which are adherent). A current theory is intervention of membrane receptors that are able to fix lymphocytes into privileged sites, particularly at the level of endothelial cells of postcapillary venules. It is interesting to note that alteration of surface receptors of thoracic duct lymphocytes, by neuraminidase or trypsin, prevents normal migration of these cells. Reappearance of the determinants is seen after incubation at 37°C, which permits the cells to once again show a specific homing capacity. A similar mechanism has been advanced for B lymphocytes that are rich in receptors for the Fe fragment of IgG and the third factor of complement. The complement system might also playa role in follicular aggregation of lymphocytes. The "homing" phenomenon, which determines the nature and organization of the peripheral lymphoid system, may thus have several mechanisms.

4

Production of Lymphoidal Cells

Traditionally, it has been a basic concept of hematology to recognize and to classify cells according to their morphology and to seek the relationship between them by considering transitional forms intermediate in morphology. Indeed, hematological diseases are recognized and classified by this very concept. Within this traditional way of thinking lie two dangers. The first is that of considering that all cells with the same name and morphology have the same life cycle, metabolic properties and function. The second hazard is assuming that the forms intermediate in morphology between any two distinct cell lines necessarily reflect a sequence of cellular transformation. A study of the quantitative aspects of lymphopoiesis as revealed by the research of the past ten years clearly shows that these aspects of the concept must be modified in order to permit the knowledge gained through basic research to be properly interpreted. From a practical point of view we must accept a morpholigical definition for lymphocytes although at the same time we must think of them as a heterogeneous category of cells. Differences in their kinetic properties and functional capacities may be used to subdivide them into groups that make up the entire category.

The problem with the morphological definition for the lymphocyte is that the lymphocyte must be identified by the absence of characteristics that other white cells possess rather than a by positive attributes of its own. A distinguishing feature of the small and medium cells is the high ratio of nucleus to cytoplasm and the clumped chromatin pattern of the smaller cells. Some of the larger cells, however, have abundant cytoplasm while other show only a small basophilic rim of cytoplasm. Furthermore, a spectrum of cells intermediate in morphology lies between these cell types. The terms large, *lymphocyte*, *lymphoblast*, and *hemocytoblast* have been applied to these cells. For the sake of simplicity, we shall refer to them as blast cells. It must be realized, however, that cells of this morphology also represent a heterogeneous category of cells with widely different functional capacities.

In seeking to establish the developmental relationship between lymphocytes and other cell lines, classical hematology relied to a great extent on the presence of intermediate morphological forms. Within the past decade, new approaches to these problems have led to an increasing realization that stem cells for the lymphoid series are a migratory population of cells of myeloid origin. These stem cells are present in myeloid tissue in both the embryonic and adult states, and from myeloid tissue they feed into the remainder of the lymphoid complex in relatively limited numbers. In lymphoid tissue, they may undergo extensive proliferation, giving rise to many generations of cells. From radioautographic considerations of the kinetics of blast cells, it is evident that cellular transformation and entry of any cell type into this category is less frequent than the occurrence of these intermediate transitional forms would suggest.

Heterogeneity of the Small Lymphocyte Population

On the basis of life span, small lymphocytes may be divided into two main groups: long-lived and short-lived. Recent evidence indicates that long-lived cells may be considered as a single group according to life cycle, kinetics, and circulation. Short lived small lymphocytes, however, are most likely not the same regarding these parameters and should not be thought of as a single group.

Long-Lived Lymphocytes

The long-lived lymphocytes comprise the majority of cells in what has been called the mobilizable lymphocyte pool or the recalculating lymphocyte pool. The experimental evidence indicates that they are the same as the thymic-dependent lymphocytes, the immunologically competent lymphocytes, and the antigen-reactive cells. All of these are new terms that have come into being in the last ten years. Thus, we will define each of them and summarize briefly the evidence that relates each to the long-lived small lymphocyte population.

The *mobilizable lymphocyte pool*, first defined by Caffrey *et al* (1962), refers to that group of small lymphocytes that may be drained

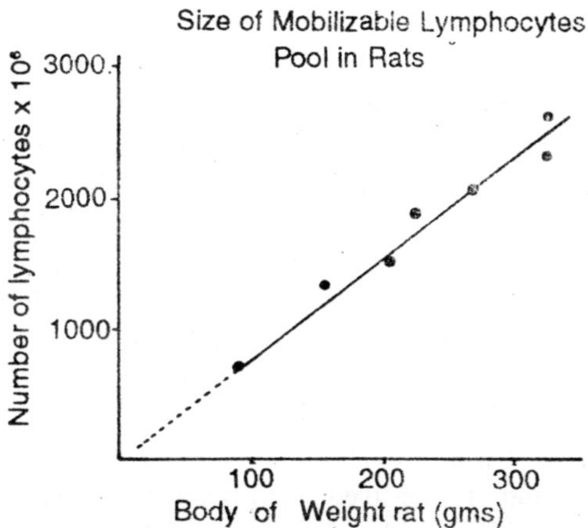

Fig. 4.1. Graph of the relationship between the number of lymphocytes that can be drained from rats with chronic thoracic duct fistulas and their body weights.

from an animal via a chronic thoracic duct fistula before a low constant output is reached. The time' usually required is two to three days of drainage. It was found that the size of the mobilizable lymphocyte pool was proportional to the body weight of the animal and contained 7.8 × 106 lymphocytes per gram for the Sprague-Dawley rat. This number represents approximately half the total number of small lymphocytes in the animal.

The *recirculating lymphocyte pool* was defined by Gowans and Knight (1964) as the population of small lymphocytes that is capable of migrating from blood to reach different lymph under normal conditions. They measured the size of the recirculating lymphocyte pool in rats by injecting labeled thoracic duct lymphocytes intravenously into syngeneic recipients and then determined the extent of their dilution by cannulating the thoracic ducts of the recipients 24 hours later. The pool of recipients' cells with which the labeled cells had mixed was estimated to contain between 1.5-2 × 109 small lymphocytes for male rats weighing 200 to 250 grams. This number is approximately the same as that obtained for the mobilizable lymphocyte pool in rats of this weight, although the two methods do not necessarily measure the same pool of cells.

Gowans (1966) has offered the criticism that the chronic drainage of lymph that measures the mobilizable lymphocyte pool may cuse a depletion of cells not normally a part of the recirculating pool On the other hand, the measurements of the recirculaing lymphocyte pool may also be criticized in that 24 hours may not be an adequate time for a complete mixing of the pool if the entire pool is not in continuous recirculation. It is conceivable that a portion of the pool may be temporarily immobilized in the lymphoid tissue.

In order to avoid these conflicts, we shall speak here of the *long-lived lymphocyte pool* and define it as that group of small lymphocytes that has a life span of more than two weeks in the rat. We shall present evidence that long-lived small lymphocytes are a single group of cells measured by their kinetics, circulation, and life cycle, and that they comprise the majority of cells in both the mobilizable lymphocyte pool and recirculating lymphocyte pool.

The first evidence comes from experiments that showed that the rate of formation of long-lived small lymphocytes was proportional to the general body growth of the rat, and that this rate may be independently calculated using the rate of disappearance of labeled small lymphocytes from spleen, lymph nodes, thoracic duct lymph, or blood in rats previously labeled with ^{3}H-thymidine during the active phase of growth. Such an observation seems highly unlikely unless the long-lived cells were in equilibrium between these compartments.

Additional evidence comes from experiments in which local irradiation was given to the cervical lymph nodes of rats in which only cells of the long-lived population were labled. The percentage of labeled lymphocytes in the locally irradiated nodes between two days and 14 days after irradiation was the same as that in the nonirradiated mesenteric nodes of the same animal and in the cervical lymph nodes of nonirradiated controls. Thus if the long-lived cells in any compartment are destroyed, that compartment is restored to normal by a migration of long-lived cells from the remainder of the pool.

Further evidence comes from experiments using rats joined in parabiosis. One member of each pair (A) was given ^{3}H-thymidine while occluding the cross circulation to the other rat (B). A variety of injection schedules was used. All labeled cells seen in the noninjected members' (B) were either immigrant cells formed in the injected members (A) or immediate progeny of immigrant cells. A comparison of the percentage of labeled monocytes or granulocytes of the injected versus the noninjected parabiont showed that these cells of blood had approximately a 25% chance of crossing from one rat to the other. However, the percentage of labeled small lymphocytes in the thoracic duct lymph was the same for both rats of all pairs irrespective of the number of 3H-thymidine injections or the time of sacrifice postthymidine.

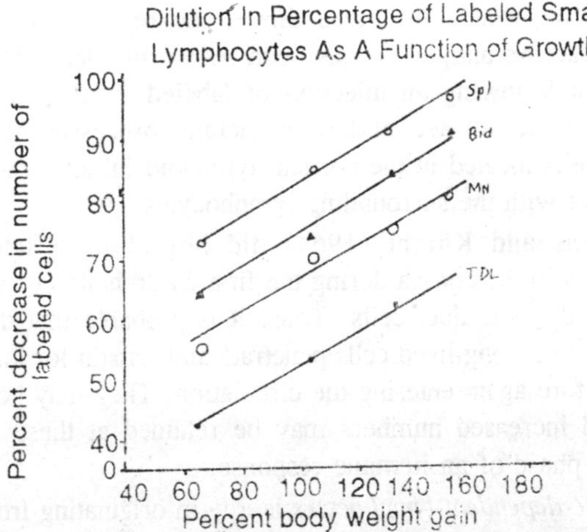

Fig. 4.2. Rats of this series received multiple injections of ³H-thymidine for two weeks during their rapid growth phase in order to label a large percentage of the long-lived lymphocytes.

This observation means that labeled cells of the recirculating lymphocyte pools were equilibrated in the two parabionts. In contrast, more than 95% of the blast cells in thoracic duct lymph of A rats were labeled, while in B rats less than 5% were labeled, showing that the blast cells do not recirculate to any great extent. Furthermore, in each pair of parabionts that was sacrificed at two weeks or more after the last ³H-thymidine injection, the distribution of labeled small lymphocytes was comparable in all lymphopoietic compartments. These observations can only mean that the great majority of the long-lived lymphocytes of lymph nodes, spleen, and Peyer's patches had entered circulation during the two-week period and were part of a single pool equilibrated by recirculation from blood to lymph.

Gowans and Knight (1964) showed that the major route of recirculation is through the mesenteric lymph nodes and Peyer's patches via the postcapillary venules. Their studies showed that labeled thoracic duct lymphocytes transfused to recipients penetrated the deep cortex of the lymph nodes and entered the medullar lymph sinuses and efferent lymphatics. However, one should not think of the long-lived small lymphocytes as being confined to this route of recirculation. The results from the parabiotic rats showed that the long-lived cells were distributed throughout the nodes and spleens with the exception of the germinal centers. However, it is of particular interest that the corona of lymphocytes surrounding the germinal centers is composed primarily of these long-lived cells, since this is the region where they most likely contact an antigen. It has been shown by electron microscopic studies that following an injection of labeled antigen, the antigen is retained on the surface of long-branching processes of specialized reticular cells located in the cortical lymphoid follicle where it makes free contact with the surrounding lymphocytes.

Gowans and Knight (1964) did not observe labeled small lymphocytes in the corona during the first 24-36 hours after transfusion of labeled thoracic duct cells. Thus, it is probable that these are the areas where the long-lived cells penetrate and remain for longer periods of time before again entering the circulation. They may home to these areas, and increased numbers may be retained at these sites during the active phase of an immune response.

Thymic dependent lymphocytes is a term originating from the work of Parrott *et al* (1966). It refers to those small lymphocytes that were found absent in specific areas of the spleen lymph nodes, and Peryer's patches in the adult animal subjected to neonatal thymectomy. The evidence indicates that these cells are the same as those of the long lived lymphocyte pool. The areas of lymphocyte depletion in neonatally thymectomized animals are the same as those primarily occupied by the long-lived cells. Furthennore, neonatally thymectomized animals have a pronounced reduction in the number of small lymphocytes in thoracic duct lymph and in their mobilizable lymphocyte pool. Labeling studies have confirmed that the reduction in the total number of lymphocytes in the thoracic duct lymph is almost entirely due to the absence of long-lived cells and that short-lived lymphocytes and large lymphocytes appear in normal numbers. However, the neonatally thymectomized animals did not show a complete absence of long-lived cells. Additional evidence has also indicated that the entire pool is not formed under.

Neonatal thymectomy does not effect a change in lymphocyte content of the bone marrow, of the red pulp of the spleen, or of the germinal centers Ricke and Schwarz (1966) found that tissue and circulating short-lived cells were present in normal numbers after neonatal thymectomy and concluded that short-lived lymphocytes are derived from lymphopoietic centers other than the thymus and that the thymus is not necessary for their formation.

Thymectomy in young adult animals also reduces the output of cells in thoracic duct lymph, although the decrease is not as pronounced as in the neonatally thymectomized animal. This evidence agrees with the concept that the long-lived pool is formed from birth to adulthood at a rate approximating that of body growth and that the formation of this pool is, mainly, thymic dependent.

The term *immunologically competent cell* was defined by Medawar (1963) as a cell fully qualified to undertake an immunological response; the term was coined by Medawar to distinguish immunological capacity from immunological performance. In contrast, an *immunologically active cell* is one already committed to and engaged in immunological activity. Three of the tests outlined for assessing immunological competence and directly related to the long-lived lymphocyte pool are:

1. the ability to restore immunological capacity to animals deprived of it,
2. the ability to produce splenomegaly, runt disease, or other manifestations of graft-versus-host reactions, and
3. the ability to produce a delayed type of inflammatory reaction by the intradermal injection of cells between animals of different antigenic makeup.

Still a fourth test that has come into being more recently is the mixed lymphocyte reaction.

An *immunologically incompetent animal* is one that does not show a normal capacity to mount a response to a large variety of antigens. In contrast, a tolerant animal is one that cannot mount a response to a specific antigen although its response to unrelated antigens appears unimpaired. Methods shown to destroy a large percentage of the long-lived pool also render the animal immunologically incompetent. Some of these are: administering high doses of total body irradiation, extracorporeal irradiation of blood, prolonged thoracic duct drainage, and treatment with antilymphocytic Serum. Neonatal thymectomy, on the other hand, renders an animal incompetent by preventing the formation of a long-lived pool of normal size. With respect to Burnet's clonal selection theory tolerance arises from the deletion or inhibition of that small number of cells in the total pool that would otherwise respond to that specific antigen. Tolerant animals have a pool of normal size, but this indicates only that the number of cells within the pool that respond to anyone antigen is extremely small. We shall discuss this point in more detail later on.

Radioautographic studies on the mechanism of action of *antilymphocytic serum* (ALS) have shown that a single injection of ALS may destroy a large percentage of the cells in the long-lived pool. In contrast, the number of short-lived cells is not reduced significantly by ALS treatment. Although the blood lymphocyte counts of ALS-treated rats may be higher than normal, lymphocyte counts of the thoracic duct lymph remain below normal for many months. Phagocytosis and destruction of the long-lived lymphocytes were evident in the liver, spleen, and lymph nodes within a few hours following an intraperitoneal injection of ALS to a rat. Macrophages laden with small lymphocytes were noted in the areas surrounding the postcapillary venues of mesenteric nodes as well as within the medullary sinuses. A single injection of ALS did not affect the lymphocyte content of thymus or of bone marrow. However, these results, it should be added, may not relate as much to specificity of ALS as they do to the life history of the long-lived pool. It might be expected, as suggested, that the long-lived cells would have more contact with ALS than the short-lived cells because of their extensive recirculation.

Immunological incompetence and tolerance should not be thought of as all-or-none phenomena: there are degrees of incompetence and tolerance. None of the methods used to render an animal incompetent completely obliterates the long-lived lymphocyte pool with the exception of lethal doses of irradiation. Also, in restoring an incompetent animal, it is not necessary to restore the entire long-live pool. Transfusion of a relatively small number of cells obtained from the thoracic duct of normal donors can break tolerance to a long-standing homograph. Shielding a single Peyer's patch during lethal irradiation protects a sufficient number of cells to cause a rejection of a homologous or heterologous bone marrow graft. Still another aspect of quantitating competence is the antigenic disparity that separates the antigens from the recipient's own tissues. Animals capable of retaining a homograft may reject a heterograft. The degree of competence is considered in transplanting homografts by doing tissue typing prior to the transplantation. Such observations indicate that the number of cells in the long-lived pool capable of responding to a specific antigen increases as the antigenic disparity separating the donor and the recipient widens. At the same time, a long-lived cell may be capable of responding to a variety of antigens. A one cell-one antigen relationship is too limiting to provide an adequate system for the innumerable natural as well as artificial antigens that exist.

The term *antigen reactive cell* was defined by Mitchell and Miller in 1968. It is the same as the *antigen sensitive cell*, a term used by others prior to that time. These terms refer to the cells that first react to antigen by entering proliferation and then initiate the differentiation of *antibody precursor cells* into *antibody forming cells*.

The evidence that Mitchell and Miller (1968) and others have presented from studies with mice is in conflict with the hypothesis that antigen reactive cells are the same as the antibody precursor cells. The antigen reactive cells appear to be equivalent to the long-lived small lymphocyte population. They have been shown to be thymic-dependent and present in the thymus in small numbers. They have been drained from normal animals by chronic thoracic duct fistula. Moreover, animals neonatally thymectomized showed a marked reduction in absolute numbers of antigen reactive cells in their mobilizable lymphocyte pools as compared to controls. Finally, antigen reactive cells were destroyed by treating animals with antilymphocytic serum. Antigen reactive cells were not found in the bone marrow, whereas antibody precursor cells were primarily of marrow origin. Thoracic duct lymph was found to contain a mixture of both antibody precursor cells and antigen reactive cells.

Short-lived Small Lymphocytes

Let us now summarize briefly what is known about short-lived lymphocytes of the different lymphopoietic compartments. Experiments measuring the rate of formation of small lymphocytes in the rat showed that the circulating life span of the short-lived cells is approximately 5 days, although for many of them it may be only a few hours. This rapid turnover rate must hold a significant meaning relating to the functions of these cells. Normally, lymphocyte destruction is most evident in the areas of tissues occupied primarily by short-lived cells (i.e., thymus, germinal centers, and bone marrow) and these are also the cells most subject to the destructive action of cortical steroids.

Thoracic duct lymph

Approximately 7% to 10% of the small lymphocytes of the thoracic duct lymph are short lived. Unlike the long-lived cells, they continue to issue from a chronic fistula in normal numbers for at least several days after the mobilizable lymphocyte pool has been drained. Neonatally thymectomized animals show a normal number of short-lived cells per volume of lymph and thus they do not appear to be thymic dependent. Since thoracic duct lymph contains antibody precursor cells as well as antigen reactive cells, and since the evidence indicates that both belong to the small lymphocyte category, it is possible that the antibody precursor cells are found within the short-lived population.

Thymus

Lymphocyte turnover in the thymus has been measured by studying the rate at which small thymocytes become labeled in the rat and mouse during the course of giving repeated injections of ^{3}H-thymidine to label all cells entering DNA synthesis. 50% of the small thymocytes were labeled in both species by 36 hours of ^3H -thymidine treatment. The rate of appearance of labeled small thymocytes is not a linear function of time but a logarithmic one, which indicates that both labeled and nonlabeled cells disappear from the thymus (or die *in situ*) at random without respect to age. Thus it was calculated that the thymus of the rat forms a volume of cells equal to the thymic weight every 2½ days.

Thymus graft experiments, local thymic labeling experiments, and experiments involving thymectomized animals agree in showing that the majority of the cells formed in the thymus die *in situ*. A small percentage (less than 5%) of those formed appear to enter the blood and course to the spleen, Peyer's patches, and lymph nodes. Localized thymic labeling and experiments in which labeled thymocytes were transfused confirm that this small percentage is a part of the long-lived pool.

The relationship between the rapid production of the short-lived thymocytes and the continued formation of the small percentage of cells leaving the thymus and contributing to the long-lived pool remains one of the most intriguing problems of lymphopoiesis. The importance of understanding this relationship is emphasized by the evidence that links both tolerance and the development of certain autoimmune diseases with the proliferation in the thymus and its contribution of cells that serves in rendering an animal immunocompetent.

Bone marrow

Bone marrow small lymphocytes of the rat have an even shorter turnover time than the thymocytes, as revealed by determining the respective rates of labeling with ^3H thymidine. It was established that their average half-time renewal rate was 24 hours. As in thymus, the rate of increase in the percentage of labeled cells was not a linear function of time but a logarithmic one and approached 100% labeling at approximately 5 days.

Local marrow labeling experiments as well as the parabiotic experiments described above showed that more than 95% of the bone marrow small lymphocytes were formed in the marrow. It has been shown that many of these enter the blood. Bone marrow lymphocytes are not immunologically competent in that they do not confer graft-versus host, reaction nor do they restore immunoiogical competence. As stated previously, the bone marrow has been shown to contain antibody precursor cells capable of coursing to the nodes and spleen and of giving rise to large blast cells that synthesize 19S antibody against sheep red cells. In view of the evidence implicating small lymphocytes as the precursors of antibody-forming cells, it is postulated that bone marrow small lymphocytes are the most likely candidates. Local marrow-labeling experiments lend further support for this postulate in that some labeled bone marrow small lymphocytes were found within the germinal centers and medullary cords of lymph nodes. Fluorescent antibody studies have established these areas as the primary sites of 19S antibody production in lymph nodes. A small number of the labeled cells of bone marrow have been reported to enter thoracic duct lymph (Brahirn and Osmond in press), and these could account for the antibody precursor cells reported to be present in lymph. Direct evidence for this is still lacking, and it remains to be shown whether the labled bone marrow lymphocytes that have coursed to spleen and nodes can transform into blast cells and whether these same blasts are the precursors to the antibody-forming cells.

It is well established that in the bird the thymus controls the formation of the long-lived pool as it does in the mammal Plasma cell production, however, appears to be dependent on another lymphoid organ, the bursa of Fabricius. The small lymphocytes of the bursa are also renewed at a rapid rate. If some of the bone marrow small lymphocytes are shown to be equivalent to the antibody precursor cells, then this population is most probably equivalent to those of the bursa.

This section of the chapter may be summarized by saying that the research of the past decade has emphasized the heterogeneity of the small lymphocyte category, and it points to the necessity of considering each group within the category as a separate population. The same research gives a strong indication that an interdependence between the various groups of small lymphocytes exists and that a traffic of cells between the hemopoietic organs is necessary to develop and maintain a normal immune system.

Stem Cells and the Interrelation of the Lymphopoietic Organs

The interrelationship of the various hemopoietic organs is further emphasized by chromosomal marker studies that have shown that stem cells for the thymocytes as well as for the long-lived lymphocytes lie within the myeloid tissue in both the embryo and adult. With respect to the thymocyte, Ford and associates (1956) identified donor type chromosomes in mitotic cells within thymuses of irradiated mice protected by bone marrow transfusion. They concluded that cells reconstituting the thymus were of donor origin. Evidence for a natural flow of stem cells into the thymus comes from experiments using parabolic mice, one member of each pair bearing a chromosomal marker. The results demonstrated an appreciable movement of cells capable of undergoing mitosis from each animal into the thymus of its partner. Taylor (1965) reported that stem cells capable of lymphoid differentiation were present in embryonic liver before their appearance in the thymus rudiment. These studies were subsequently confirmed by Owen and Ritter (1969) and by Auerbach and Globerson (1966). Their chromosomal marker studies clearly showed that the lymphoid elements of the thymus originated in the embryonic liver.

The precursors of the immunologically competent cells have also been shown to be of myeloid origin and the formation of these cells was found dependent on an interaction of bone marrow and thymus. Tyan and Cole (1966) have used chromosomal marker studies to show that potentially immunologically compctent cells derived from the mouse fetal liver migrate into the thymic tissue where they mature and/or proliferate prior to their appearance in peripheral lymphoid tissue. Using organ cultures, Auerbach (1966) reported that although cells within embryonic liver where not immunologically competent as measured by their ability to produce graft-versus-host reactions, competence was acquired when the liver explants were cultured for several days in combination with thymic tissue. Furthermore, when spleen rudiments were obtained from the embryo and cultured, in a similar manner, no lymphoid development was observed. If a spleen graft was made in the anterior chamber of the adult mouse eye, lymphoid development was found and shown to be the result of cell migration. Auerbach concluded that the thymus bone marrow and spleen all participate in some interdependent way in the development of an immunologically competent animal McGregor (1968) has presented evidence that the immunologically competent cells of thoracic duct lymph develop from myeloid tissue. In adult rats inoculated at birth with parental strainbown marrow cell, the choracic duct lymph was shown to contain cells having the immunological capacity of the bone marrow donor.

These experiments are only a few that emphasize that the lymphoid stem cells are migrating population in both the embryonic and adult state, and that the development of a normal lymphoid complex depends on the interaction of cells from different organs. However, one must be aware that none of these experiments depends on a massive flow of cells from one organ to another. In a normal animal, labeled transfused cells of bone marrow, spleen, node or thymus origin do not home to the thymus in sufficient numbers to be recovered within the thymus of the recipient. Although there was a good exchange of labeled blood cells of all types in the studies employing normal parabiotic rats, less than one cell in ten thousand thymocytes was labeled in the noninjected member, whereas essentially all of the thymocytes were labeled in the members injected with ^{3}H -thymidine. The majority of the immigrant cells to the thymus had the morphology of a monocyte and more rarely a labeled small lymphocyte was encountered in the medullary areas. These observations simply mean that the absolute number of stem cells entering the thymus is small when compared to the number of large thymocytes present, and that the transformation of a stem cell into a blast is an event too rare to be observed in the normal thymus. The results of experiments in which a thymus bearing chromosomal markers was grafted to a normal recipient support this concept in that it takes from two to three weeks for the donor mitotic figures to be replaced by recipient cells. Thus the thymus has within it cells capable of giving rise to many generations of progeny, although the ability for self-renewal appears to be limited. The failure to recognize the possibility that stem cells may exist in such limited numbers that they may go undetected morphologically is perhaps the greatest criticism to most studies that in the past have relied on morphological criteria only for studying cellular derivation and differentiation. One of the greatest challenges for the future is to devise and employ new techniques that overcome this criticism.

The morphology of the stem cells for the thymus has been a subject of intensive interest in our laboratories in the last few years. Under conditions afforded by irradiation, in which the thymus was first depleted, of the lymphoid population, it was possible to show a migration of labeled cells of marrow origin into the thymus. In parabiotic experiments employing irradiation and marrow shielding, and in experiments involving the transfusion of labeled marrow cells to irradiated rats, the only labeled cells migrating into the thymus were similar in morphology to blood monocytes. We have called these "monocytoid cells," since it is realized that all cells that look alike do not necessarily have the same functional capacities or metabolic properties.

Studies on the quantitative aspects of stem cells in peripheral blood and bone marrow clearly indicate that the stem cells capable of self renewal and the extensive proliferation necessary for reseeding a lethally irradiated animal are present in very limited numbers. Lewis and Trobaugh (1964), found five spleen colonies for every 10,000 injected marrow cells and five colonies for every 500,000 to one million injected blood leukocytes. Although the spleen-colony method does not assess the absolute number of stem cells in any transfuse, it should be obvious from such calculations that the changes of isolating a stem cell allowing for morphological characterization are remote. Nevertheless, the only cell type that entered the thymus had abundant pale staining cytoplasm a nucleus with many invaginations and folds and fine chromatin structure. The majority of these labeled cells entered the thymus of the irradiated rat via the blood vessels within the septa, and made their way through the connective tissue to the outer cortex. Labeled mitotic figures were seen and clusters of weakly labeled blasts were encountered at 24 to 48 hours posttransfusion.

In 1966 we presented radioautographic evidence in conflict with the classical concept of hematology, which advocates that cells of the fixed reticular network of the hemopoietic organs give rise to the free cells therein. The stromal cells of the spleen, thymus, lymph nodes and bone marrow were shown to be proliferating at a very slow rate. A small percentage was found to incorporate ^{3}H -thymidine and give rise to themselves. A larger percentage were labeled by multiple injections of ^{3}H -thymidine to growing animals, and a slow turnover time for reticular cells was confirmed by experiments measuring the disappearance rate for labeled cells after long postthymidine intervals. Many reticular cells were still labeled in rats sacrificed at one-year post-^{3}H -thymidine. It may be noted that stromal cells of the hemopoietic organs have long been recognized as being highly radioresistant. Thus, sublethal irradiation experiments were designed to stimulate blast -cell formation and to test the capacity of the labeled reticular cells to transform. No evidence was obtained to support the view that reticular cells have stem cell capacity. The more recent evidence showing that the stem cells for both the short- and long-lived populations are of myeloid origin are a highly mobile population, and are radiosensitive further undermines this classical concept.

Small Lymphocyte Transformation

The results of radioautographic studies clearly show that the frequency with which a long-lived small lymphocyte transform to blast cell *in vivo* is infrequent. Rats sacrificed at two weeks or more after multiple injections of ^{3}H -thymidine and which had as many as 50% of the long-lived lymphocytes labeled showed less than 1:10,000 blast cells to be labeled as calculated from radioautographs of tissue sections.

Within the first 24 hours after antigenic stimulation with normal rabbit serum, this frequency was increased, in nodes and spleen, but it still represented less than 1:1,000 blast cells.

Miller *et al* (1967) determined the number of antigen reactive cells within the mobilizable lymphocyte pool of mice that responded to sheep erythrocytes. They reported that approximately 60 cells per million lymphocytes responded.

Studies employing radioautography as a means of estimating the number of small lymphocytes that transform in the *in vitro* culture systems also show that relatively small percentage of the lymphocytes respond. At least three independent studies have shown that less than 3% of the original small lymphocyte population respond in systems measuring immunological reactions. These approaches have included the monolayer culture system, analogous to the homograft reaction, the mixed lymphocye reaction, and the incubation of lymphocytes with antigens such as tuberculin.

More recently, Marshall *et al* (1969) have followed the proliferation of single blasts for three to five days of culture using time-lapse cinephotomicrography. They selected lymphoblasts derived from 48-hour cultures with tuberculin, pokeweed, or mixed lymphocyte cultures, and at that time the number of lymphoblast was approximately 2.3% of the initial small lymphocyte count. Examination of the film showed that the lymphoblasts divided and redivided to produce clones of 64 cells or more with an average generation time of eight to 13 hours.

The precursors to the antibody-forming cells also appear to respond in very small numbers to anyone antigen as assayed by the plague-forming technique. Mitotic blocking agents such as colchicine or velban have been used in this system to inhibit the proliferation of antibody-forming cells in order to assess recruitment of antibody-forming cells. It was concluded that following antigeneic stimulation with sheep red cells, relatively few cells responded by synthesizing antibody and that these in turn entered a proliferative sequence. The magnitude of the antibody response was dependent primarily on the rate of division of antibody producing cells and not on new cell recruitment.

Blast Cell Formation and Kinetics

If one accepts the evidence showing that stem cells comprise only a minute fraction of the cells in myeloid tissue and that the transformation of the small lymphocyte is a rare event in the nodes or spleen of the intact animal, then we are faced with the problem of what is the precursor to the relatively large number of blast cells that are readily observed. In order to answer this question, it is necessary to investigate the kinetics of blast cells.

Early hematologists realized that the cells that they referred to as *hemocytoblasts*, and were found in the lymph nodes, spleen, bone marrow, and thymus, had a short generation time as indicated by their mitotic index. Radioautographic studies have confirmed this concept. The average blast cells in lymph nodes and spleen have a generation time of 12 hours, and in thymus it is approximately 9½hours.

Mesenteric lymph nodes from rats sacrificed within 15 minutes of a single injection of ^{3}H -thymidine had approximately 50% of the blasts cells labeled. Thus 50% of the blasts were in DNA synthesis at anyone time. Rats sacrificed at 12, 24 and 36 hours after a single injection also had 50% of the blast cells labeled but the average grain count had decreased as a logarithmic function of time. The half time was 12 hours. These results show that the majority of the blast cells in the node are derived from themselves and that they represent successive generations of cells with the same morphological appearance. Although it is possible that the enlargement of a small lymphocyte or the transformation of a yet unidentified stem cell initiates differentiation, quantitatively speaking, and cell reproduction by the blast cells themselves overwhelmingly outweigh the event of cell transformation.

This principal of differentiation is further illustrated by the bone marrow. Cells that have morphology similar to the blast cells and lymph nodes and thymus are also present in bone marrow. The large and medium forms have been termed *transitional cells* by Yoffey (1962), who has considered them intermediate between .the small lymphocytes of the bone marrow and the cells of the erythrocytic and granulocytic series. The local marrow-labeling experiments which showed that small lymphocytes are formed in the marrow, also gave evidence that at least some of the transitional cells gave rise to the marrow small lymhocytes. Recently, the transitional cells have been divided into two morphological types pale and basophilic. Studies of their kinetic behavior show that the majority of the cells in each category give rise to themselves for many generations. Although some indirect evidence suggests that precursors to the proerythroblasts may be found in the transitional cell category, this question must remain open.

Heterogemeoty of Lymphocytes

The lymphocytes constitute a family of cells of different origins, migration patterns, sizes, staining characteristics, ultrastructure, life span and function. The smaller lymphocytes have a small, dense nucleus in which nucleoli are rarely visible; the larger ones have a round or slightly indented nucleus with a loose chromatin structure, and nucleoli are always visible with proper staining techniques. Because the cytoplasm in lymphocytes of the smaller class is generally very pale, usually only, a small rim on one side of the nucleus is visible; electron microscopy shows that these cells have very few single ribosomes and no endoplasmic reticulum. However, some small lymphocytes presumably contain more ribosomes, since the cytoplasm is much more basophilic and stains intensely with pyronin. The larger lymphocytes have more numerous single ribosomes and in some instances, such as following the action of antigenic stimuli, one can observe polyribosomes with a faint endoplasmic reticulum in cells that may be in the process of transforming into plasma cells. Bovine thoracic duct lymphocytes have been shown to be of two overlapping size populations. The population of smaller cells has a median volume of 250 cubic microns with a 99 percent confidence range from 142 to 425 cubic microns, while the population of larger cells has a median volume of 648 cubic microns with a 99 percent confidence range of 370 to 1,150 cubic microns. The populations intersects at 312 cubic microns. For practical purposes, the cells with a volume of less than 312 cubic microns are considered to be the small lymphocytes, and those with a volume greater than 312 cubic microns the large.

The life span of lymphocytes has been reported to vary from a few hours to years. Little and associates (1962) have shown that the smaller class of rat lymphocytes is divided into those with relatively short life spans, and those with life spans greater than one year. The larger lymphocytes are also divided into two components; one with a life span of the order of one to two days, and the other with a maximum life span of about 60 days. Norman and associates (1965) have demonstrated that small lymphocytes have a mean life span of 530 ± 64 days in women.

On the basis of a study of mitotic indices and labeling with tritiated thymidine (^{3}HTdR), Osogoe (1963) concluded that there are two populations of lymphocytes; the first is produced rapidly, homes of the gastrointestinal tract and has a short life span; the other is smaller in size and has a very long life span. Everett *et al* (1964) also utilized single and multiple injections of ^{3}HTdR and analyzed the behavior of lymphocytes in blood, thoracic duct lymph, and diverse organs. They concluded that 90 percent of the cells in the thoracic duct lymph are long-lived. The long-lived cells are produced at a. rate proportional to body growth and they recirculate from blood lymph. They also deduced that the thymus may be a source of the long-lived circulating lymphocytes. The small lymphocytes in the bone marrow, approximately 95 percent of those in the thymus, and a major percentage of those in the spleen, are short lived (five days or less although it is nuclear whether these cells arise from the cortex, or the medulla, of the thymus. Long-lived lymphocytes are immunologically committed and serve as a basis for immunologic memory.

Continuous intravenous administration of ^{3}HTdR in rats showed that the median survival time of small lymphocytes was about one month in animals that had received ^{3}HTdR for up to 271 days while 5% to 8% had a life span of more than nine months. Large lymphocytes constituted a homogeneous population in respect to labeling but small lymphocytes labeled nonuniformily, and appeared to comprise at least two populations differing in intensity of labeling and turnover rates with the more heavily labeled cells having the faster turnover.

Generate Cycle of Lymphocytes and the Rate of Proliferation

Determining the time a cells spends in each phase of the generative cycle necessitates either direct observation or the introduction of a label at one of the subdivisions of the generative cycle; for example, by labeling all cells in the phase of DNA synthesis with ^{3}HTdR. Thereafter, one can observe changes in the fraction of cells that are labeled and in the flow of labeled cells through the premitotic rest periods and through mitosis. In the lymphatic nodules of the rat spleen, the autoradiographic data suggested a DNA synthesis time of about five hours and a minimum time for prernitotic rest of 30 minutes, with a maximum between one and two hours. The diminution in intensity of label over interphase cells indicated a half-time of 13.4 hours but the grain count over mitotic figures showed a half-time of six to seven hours, which probably approximates the true generation time; this difference is probably an indication of the reutilization of tritium labeled breakdown products of DNA. There is a great contrast in the rate of proliferation between the mantle-zone cells and the germinal-center cells. One to two per cent of mantle-zone cells surrounding the germinal center becomes labeled after a single ^{3}HTdR injection. The grain count data over these labeled cells are compatible with the idea that the cells undergo one division within the first twelve hours after injection of tritiated thymidine. Thereafter, further divisions may not occur for days and a half time for the diminution in intensity of 260 hours or more may be calculated. This suggests that 98% of the mantle cells are in G_0 or are resting with a long G_1.

Various phases in the generative cycle of thoracic duct cells in calves have been studied. These results showed that the generative time of the more basophilic cells is about 5½ to 6 hours, with a DNA synthesis time, of 4 hours, while the populations of less basophilic cells had cycles about one to two hours longer. Mitotic time was found to be about 18 minutes, which is substantially shorter than the 40 minutes obtained by direct observation of an antibody producing cell in mitosis. Plasmablasts and hemohistioblasts in mesenteric lymph nodes of the rat have generative times of about 9 and 12 hours respectively. Mature plasma cells probably constitute a nondividing population, which is renewed in the lymph node in not more than five days.

Mouse spleen cells had a doubling time of 24 hours during the latent and log phases of primary antibody response, which decreased to 14 hours after antigenic stimulation. The doubling time can be assumed equal to the generation time only if all labeled cells divide and if there is no death of the progeny. There seems little doubt that antigens shorten the doubling time to a certain extent and stimulate immature plasmalike cells and lymphocytes to enter the cycle. The cell cycle of blast cells derived from spleens of primed donors in which a secondary response was initiated in diffusion chambers by the addition of antigen, showed a generation time of about 8 to 9 hours during the latent to log phase transition through the first mitosis, a mitotic time of 0.5 hours and a presynthesis rest of 0-1 hours. The short synthesis and generation times have been confirmed by Vincent *et al* (1969), using double labeling techniques.

Quantitative studies on lymphocyte production were begun by Yoffey (1936), but the initial measurements were high because of failure to consider the recirculaticn of lymphocytes from blood. Counts of mitotic figures throughout the lymph nodes, the thymus, and the spleen of the rat, assuming a one-hour mitotic time, show that approximately 67.6 $\times 10^6$ lymphocytes are produced per day per 100 grams of body weight. If one reduces the mitotic time to the 18 minutes measured by Cunningham *et al* (1967), the production rate can be increased by a factor of about 3 or 4. About 0.5% to 2% of the DNA of rodent lymphoid tissue is renewed per hour in the lymph nodes and the thymus. Conversion of the DNA renewal rate into number of cells produced per unit of time has been done but is fraught with potential errors, such as the reutilization of labeled material from the death of cells and variations in size of metabolic pools. Therefore, by this method it is difficult to determine the total number of cells that are actually leaving the organs. Cronkite *et al* (1964) showed that the output in the thoracic duct of calves was about 2×10^8 per minute of which 10 per cent are newly produced and the others recycling. It was further estimated that four to 12 lymphocytes entered the blood from some other sources for every cell that entered the blood from some other source for every cell that entered through the thoracic duct. Single, normal, axillary lymph nodes produce about 0.5×10^6 cells per hour, but after antigenic stimulation by skin allograft this increase to 2.6×10^6, concluding that the increased production of cells is due to an increase in the number of cells replicating rather than to a decrease in the generation time. Their estimated generation time of nine hours is close to the value observed by Fliedner *et al* (1964) and Janeit *et al* (1966).

While it is simple to measure the number of cells coming out of the thoracic duct or an efferent lymphatic, it is still very difficult to get any realistic estimate of the absolute production rate of new lymphocytes in all the different morphologic categories. There are uncertainties in determining the migration pattern, the fraction of cells that are recycling from blood to lymph, the reutilization of DNA labeling materials, and the possible increased death of cells by antigenic or other stimuli; but it appears to be clearly established that antigens are effective in increasing the rate of proliferation.

Under normal conditions, the mass of the adult lymphoreticular system is relatively constant, implying a balance between the production and death of cells that may arise from a small constant input of antigens from the gastrointestinal tract or by absorption through the skin. One can determine the number of newly produced lymphocytes that leave a single lymph node by combining efferent duct cannulation with arterial ^{3}HTdR perfusion in order to label the DNA of all new cells. However, the absolute production can be determined from the replacement of unlabeled by labeled DNA per unit time divided by the average DNA per cell, plus the number of labeled cells leaving the node. The replacement of unlabeled by labeled DNA in the steady state represents the intranodal cellular death rate of cells produced in the node. Imperfect knowledge of lymphopoiesis prevents accurate assessment of the absolute production rates at this time.

The lymphocytes are also heterogeneous in respect to their organ of origin-namely, the thymus, lymph nodes, spleen, gut-associated lymphoid tissue, or the bone marrow and to their function. Various investigators have ascribed essential roles in primary immunity, secondary immunity, allograft rejection, delayed hypersensitivity, tolerance, transfer reaction such as graft us host disease, as stem cells for hemopoiesis in general and in tropic functions. Finally, lymphocytes are heterogeneous in respect to their content of IgG, IgM, IgA, and antibodies; it is unusual, but not unknown, to find a cell with more than one Ig or antibody.

Migration Pathways

The flow of lymphocytes from the bone marrow to the blood and from blood to marrow has been studied over the years by many investigators. The concept that there is a major migration stream of small lymphocytes from the blood to the marrow accounting for the daily replacement factor, and that the migrant lymphocytes are transformed and serve as stem cells is not now accepted. In the dog, ^{3}HTdR-tagged small lymphocytes labeled at large in the body migrated to the bone marrow, but there was no evidence of transformation into erythroblasts. The evidence suggested that there are two populations of small lymphocytes, one that migrates to and one that is generated within the marrow. There is strong evidence that some of the numerous small lymphocytes found in guinea pig bone marrow; are rapidly renewed by proliferation of precursor cells within the marrow. In these shldies, Osmond and Everett used, in part, techniques of excluding circulation to marrow while ^{3}HTdR is administered, similar to those of Keiser *et al* (1964) in the dog. In addition, they studied the behavior

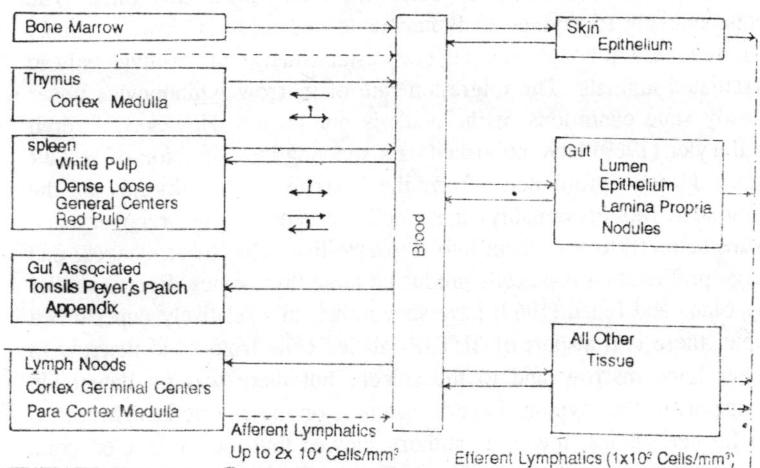

Fig. 4.3. The ubiquitous lymphocyte, its exchanges with blood and tissues.

of marrow small lymphocytes in diffusion chambers where there can be no migration of lymphocytes into the chamber. They concluded that there is no large-scale migration of small lymphocytes into the marrow but some migration of large undifferentiated cells indicating intramedullary proliferation of precursor cells to form the marrow small lymphocytes. Rat studies also indicate that the marrow is a major site of proliferation of small lymphocytes. Although the evidence for proliferation of small lymphocytes in the marrow is secure, the nature of the precursor cell is not clear. It is conceivable that the small lymphocyte is in a long G_1 or G_0 phase that becomes the transitional cell of Yoffey (1958) after stimulation. The genetic relationship between the large and small-marrow lymphocytes remains to be determined. Further evidence for increased intramedullary production of lymphocytes by hypoxia confirms earlier comparable studies in guinea pig and mouse.

Since there is a large-scale intramedullary production of lymphocytes, as well as some migration of small lymphocytes into the marrow, steady state conditions demand a large-scale intramedullary death or migration out of the marrow. Hudson and Yoffey (1966) have demonstrated, by electron microscopy, that lymphocytes migrate between the sinusoidal endothelial cells of the marrow but not through them, and the direction of movement could not be determined. Syngeneic marrow transplantation using chromosome markers has shown that the marrow contains cells capable of repopulating the thymus, spleen lymph nodes, and bone marrow of otherwise fatally irradiated mice. The bone marrow to thymus to lymph node migration is slow, probably taking weeks although this has been established with certainty only in irradiated animals. The migration rate of marrow lymphocytes under steady state conditions in the adult is not known. However, Everett and Tyler (1969) have published data showing the migration of ^{3}HTdR labeled monocytoid cells from the marrow to the thymus of the irradiated rat. Presumably, these cells promote thymic recovery and were believed to transform into fibrocyte-like cells, macrophages, and large proliferating blast cells producing large thymocytes. Recent studies by Linna and Liden (1969) have shown that, in a relatively unperturbed state, there is transport of 3HTdR labeled cells from bone marrow to other bone marrow and to the spleen, but there was no detectable transport to the thymus, Peyer's patches, or mesenteric lymph nodes. In further studies, the same authors showed that locally labeled bone marrow lymphocytes migrated to lymph nodes and skin areas after delayed type sensitivity, but there was no preferential migration to the nodes regional to the skin sensitization. The cells were found in the cortex, medulla, and lymphocytic collars around germinal centers but not in germinal centers. The authors interpreted this migration as being nonspecific, since cells were found in lymph nodes other than those regional to areas of skin sensitization. The observations are pertinent to the suggestion that bone marrow-derived cells are precursors of antibody-synthesizing-cells and that thymicderived cells are antigenreactive-cells, which was supported by the observations that both 19S hemolysin-producing cells and immuno-competent cells in the graft-versushost reactions were marrow derived.

In a recent study, cells in neonatal marrow were labeled by continuous. ^{3}HTdR infusion throughout pregnancy. The labeling of bone marrow lymphocytes dropped rapidly during the first week of life, as did the grain count overlying the cells, but both remained constant for 21 weeks thereafter. This suggests that there is a long-lived and short-lived marrow lymphocyte and may indicate migration out of the marrow, but does not provide information on the rate of migration.

Thymic Migration Routes

The apparent high production rate of cells in the thymus must be balanced by death or migration of cells or there will be progressive growth of the thymus. Dye exclusion studies indicated that 3% to 5% of mouse thymic cells were dying (stained) in the first three months while in from 3 to 12 months the dying cells increased to 15% to 20%. In foetal life, the death rate was higher on the sixteenth day of gestation than on the eighteenth. Logically, the lower death rates can be correlated with periods of higher migration rates. The question of migration versus *in situ* death of lymphocytes in the thymus in newborn mice has been studied by simultaneously measuring thymic growth and cell production in the thymus, and it was found that one-third of all cells present in the thymus were lost during the third day of life. In view of the low pyknotic index and Claesson's (1969) observation of low death rate discussed above, an extensive migration must have taken place during the third day. In view of the preceding, the earlier notion that 99%, of thymic lymphocytes disintegrate within the thymus must be rejected for the time intervals studied. In young adult mice, it has been shown that cells migrate from the outer cortical zones into the perivascular lymphatic vessels near the corticomedullary junction; and that a large fraction of lymphocytes leave the thymus through lymphatics. There is also a series of papers suggesting that lymphocytes leave the thymus via the veins. Thus, the evidence for migration out of the thymus during fetal, neonatal and young adult life is unequivocal. The magnitude of the migration varies with the age of the animals, but its significance is not clear. Some insight into its possible importance can be gleaned from the following studies.

Neonatal thymectomy in mice leads to severe lymphocyte depletion from specific areas of the spleen and lymph node defined as *thymus-dependent areas*; somewhat similar observation had been made earlier in the rat. Extracorporeal irradiation of the blood produces a depletion of lymphocytes in the paracortical areas of the lymph nodes, the cuff of small lymphocytes surrounding germinal centers in the spleen and lymph nodes, and the loose white pulp in the spleen. These areas of the spleen and lymph nodes are considered thymic-dependent. The proliferation of pyroninophilic blast cells, which accompanies cell-mediated immune reactions, has been shown to be confined to the "thymus-dependent" or paracortical areas of lymph nodes of guinea pigs and mice.

Syngeneic thymus cells, which have been labeled *in vitro* with tritiated adenosine, localize preferentially in the "thymus-dependent areas" after intravenous injection into mice. Mice with congenital aphasia of the thymus show marked depletion of lymphocytes from the "thymic dependent areas" of the lymphoid organs. Goldschneider and McGregor (1968) labeled syngeneic rat thymocytes *in vitro* with ^{3}H-5-uridine and found similar pathways in normal and neonatal thymectomized rats and in rats depleted by thoratic dust drainage. Both large and small thymocytes filtered through lung more slowly than thoracic duct lymphocytes and accumulated in liver sinusoids. Large thymocytes entered spleen red pulp and the intestinal wall while small thymocytes entered spleen white pulp along the same route as small lymphocytes and penetrated into the diffuse lymphocytic fields of lymph nodes and Peyer's patches *via* post capillary venules. A few small thymocytes were found in the thymus and bone marrow. Large thymocytes were labeled by a single intravenous injection of ^{3}HTdR thymic lymphocyte suspensions were injected IV. The large thymocytes were not found in lymph nodes, thymus, or Peyer's patches, but were found in lung, liver, spleen red pulp, intestinal wall and, occasionally, in spleen white pulp. This approach, like all techniques involving the administration of thymic lymphocyte suspensions, is open to the criticism that some of the cells in such suspensions would not normally leave the thymus. Studies with thymus grafts or with chromosome markers have demonstrated the migration of thymic lymphocytes to the spleen and lymph nodes of mice, although whether this represents normal migration is still unresolved.

In situ labeling with ^{3}HTdR is preferable since it labels only cells in DNA synthesis using a single slow intraarterial injection via the thymic artery or direct injection in the thymic gland. Subcapsular injection may cause trauma to the thymus and its lymphatics while injection directly into the thymic artery may damage this vessel and cause interference to the normal flow of arterial blood into the injected thymus.

Sicne we were unsatisfied with the existing techniques, we developed a technique for long-term intraarterial ^{3}HTdR perfusion of the thymus of calf. In our studies, the perfused thymic weight is 20 to 83 grams or about 20% to 40% of the whole thymus. Studies on anatomic distribution of cells originating in the thymus of calves killed after two to eight days of infusion have shown that the majority of heavily labeled cells appeared in the "thymic-dependent areas" of the spleen and lymph nodes. The density was greater in the spleen, but few heavily labeled cells were seen in or near post capillary venules. Thymic cells were also found in the red pulp of the spleen, medulla, and the outer cortex of the lymph node. No heavily labeled cells were

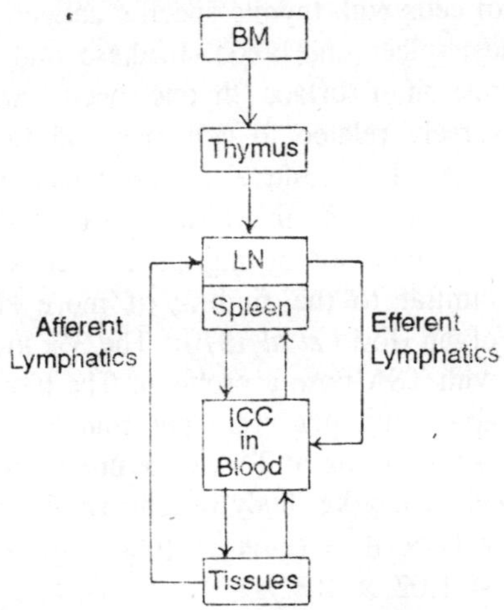

Slwo Bone Marrow Lymphocyte Migration
During Regeneration

Fig. 4.4. Schematic presentation of the slow lymphocyte migration demonstrated during regeneration of hemopoietic tissues in otherwise fatally criadiated mice saved by syngeneic bone marrow transplantation.

found in the bone marrow but large fractions of cells in Peyer's patches were lightly labeled. In other unpublished studies, it appeared that an occasional thymic lymphocyte found its way back to the contralateral thymus at a region of presumed exit (corticomedullary junction), suggesting that recycling may also occur in the thymus. Some heavily labeled cells are found in the thoracic duct, indicating that thymic cells enter the recycling pool. However, the thymic cells appear to be short-lived or divide rapidly since none were found in the tissues 96 hours after termination of perfusion.

Our suggestion of migration of cells to the thymus is supported by observed changes in specific activity of DNA after administration of ^{3}HTdR, irradiation and chilling. The slow regeneration of bone marrow, thymus and spleen and lymph nodes after marrow transfusion in fatally irradiated mice is well known and is shown schematically.. The role of this route in reestablishing immunocompetence has been generally accepted but the effect of the apparent large scale migration of cells to thymus in the studies of Field and Stanley (1966) is not clear, although it could represent migration from marrow to thymus to produce antigen-reactive cells.

Another approach that we have used to study thymic cell migration is the observation of cells with thymic specific antigen in the thymus, lymph, lymph nodes, spleen and blood. In these studies, it has been shown that the expression of surface, thymic-specific antigen (TSA) in the thymus, is inversely related to cell size and to expression of histocompatibility antigen. TSA-positive cells are found in thymic lymph, thymic vein, lymph nodes, spleen arterial blood, and thoracic duct lymph. The highest concentration of migrant TSA cells is found in the spleen, which is similar to the finding of more ^{3}HTdR labeled thymocytes in that organ (Iorio *et al* 1970). The 3% to 8% of cells in thymic vein blood with TSA pose a problem. The blood flow rate in bovine thymus is about 1ml per gram per minute. If each cell in thymic vein blood with a coating of TSA were produced in the thymus, the number produced per day/kg. body weight would be:

$A \times B \times C \times D \times E = (1.44 \times 10^3) \times (6 \times 10^6) \times (6 \times 10^{-2}) \times 1.97 \times 1 = 1.02 \times 10^9$

A = minutes per day,

B = concentration of lymphocytes in thymic vein blood per ml,

C = fraction with thymus specific antigen in thymic vein blood minus fraction in arterial blood,

D = weight of thymus in grams per kg body weight = 1.97,

E = blood flow (in ml) per gram per minute through thymus.

Since the blood of calves contains about 0.3 x 109 (smalilymphocytes)/kg, the daily output from the thymus is nearly four times greater than the number of small lymphocytes in the blood. The magnitude of migration is perhaps conceivable and would certainly explain the reduction in blood lymphocyte concentration and the reduced thoracic duct output following thymectomy, but one must also consider the possibility that blood lymphocytes enter the thymus and pick up a coating of thymic specific antigen. Emstrom and Larsson (1967) have also published evidence that the guinea pig thymus produces enough lymphocytes to replace blood lymphocytes 3.7 times per day, a number very close to our estimate in the calf.

Splenic Migration

There is considerable recent literature on the flow of lymphocytes into, and out of, the spleen based on arteriovenous differences in lymphocyte concentration, which indicate a substantially greater concentration of lymphocytes in the splenic vein, suggesting a net input into the blood. Ford (1969a) has demonstrated that lymphocytes recycle through the spleen at a rate proportional to their concentration in blood; the minimum transit time through the spleen is about two to three hours. The route of migration involves the trapping of lymphocytes in the periarteriolar sheath of the marginal zone and entering the

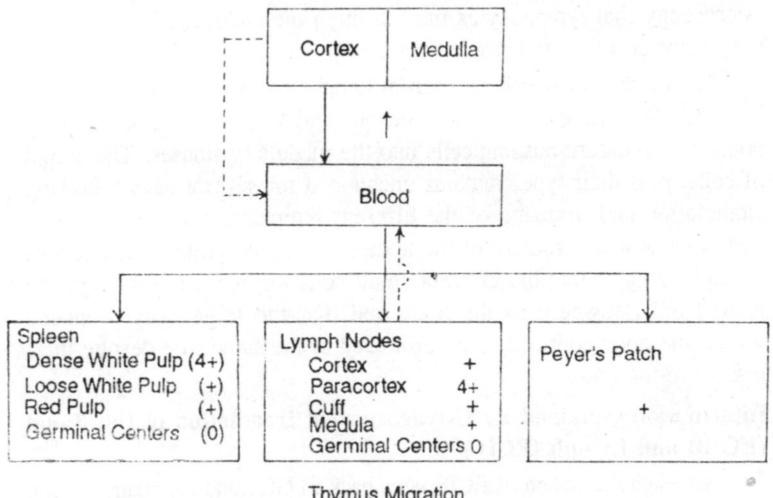

Fig. 4.5. Schematic routes of migration of thymic cells showing sites of relative concentrations.

white pulp, as previously suggested. The elegant morphological and ultrastructural studies of the spleen by Weiss (1962, 1963) show that tiny arterioles, which leave the central arteriole and terminate in the white pulp, rarely contain red cells but do contain lymphocytes. One presumes that plasma with lymphocytes is skimmed from the flowing column of blood and that lymphocytes are deposited in the marginal zone, thus providing a mechanism for the observed anatomic distribution, from which they make their way into the red pulp and presumably back into the blood. Ford (1969b) has also shown that cells recirculating through the thoracic duct can recycle through the lymph nodes; cells in both tissues have similar immunological properties.

Lymph Node Recycling

Recycling is so well known now that discussion can be brief. The site of recycling is the postcapillary venial, known for years. The notion of recycling of lymphocytes from blood to lymph apparently was initiated by Renaut (1888) and again advanced by Sjovall (1936). In a series of elegant experiments Gowans (1959) and Gowans and Knight (1963) labeled syngeneic rat thoracic duct lymphocytes with tritiated adenosine, an RNA label, and conclusively demonstrated the recycling of lymphocytes from the thoracic duct to the blood and then through the postcapillary venules of the lymph nodes and back into the thoracic duct. Marchesi and Gowans (1964) showed by electron microscopy that lymphocytes pass through the endothelial cells rather than between them as granulocytes and monocytes do.

The outflow of lymphocytes from lymph nodes in efferent lymphatics indicates that there is a passive desquamation of cells apparently by population pressure pushing cells into the medullary sinuses. The output of cells, plus their type, remains unchanged for several days following cannulation and drainage of the efferent lymphatic of a lymph node, and the anatomic structure of the node remains the same. These results strongly suggest that the efferent small cells are recycling through the node from elsewhere in the body and that the large cells produced within the node continue to be produced at the same rate despite their loss from the body.

Information Obtained by Extracorporeal Irradiation of the Blood (ECIB) and Lymph (ECIL)

Although the notion of ECIB goes back to Gley and Heymans (1921), it was perfected as a tool for study of lymphopoiesis by Cronkite and his associates, who took advantage of the relative radioresistance of erythrocytes and the radiosensitivity of lymphocytes. Killing the blood lymphocytes should result in depletion of the lymphoreticular tissues in general since the blood lymphocytes exchange freely with some of the lymphocytes of lymphoreticular tissues. Irradiation of the blood and/or of the lymph has become a simple technique utilized in many laboratories. It has been useful in the induction of lymphopenia, depletion of small lymphocytes from lymphoreticular tissues, modifying allograft reactions and to some extent in the treatment of animal and human leukemia.

Depletion of the Lymphoreticular Tissues by ECIB

During ECIB, the thymic cortex is diminished in thickness, the cortiocomedullary ratio is decreased, and the spleenic lymph follicles are markedly decreased in size, as a result of loss of small lymphocytes from the cuff surrounding the germinal centers. There is also a remarkable loss of lymphocoytes from the less densely populated areas of white pulp, as well as from the red pulp. The germinal centers in the spleen and lymph nodes are intact and have an increase in mitotic activity. There is a striking diminution in the thickness of the cortical lymphocyte population in the lymph nodes, and in the relative density of lymphocytes in the medulla. Disintegrating cells are found in the intrafollicular zones of the cortex, or near the corticomedullary junction. The quantitative aspect of the depletion of lymphoreticular tissues with time have shown that all of them and the blood have a two-component depletion during 48 hours of ECIB. Approximately 50 percent of the small lymphocytes in the cortex of the lymph nodes are tightly fixed and are mobilized very little, if at all, constituting a slowly mobilizable pool. Evidence for a mobilizable lymphocyte pool of 7.8×10^9 cells per kg body weight has been obtained from studies on thoracic duct drainage in the rat. The anatomic site of this was not determined as it was in the studies of Ruchti *et al* (1970).

Size of the Lymphocyte Pools Accessible by Thoracic Duct Drainage

The concentration of small lymphocytes in the peripheral blood and the output/ kg / hr has been studied during 23 days of continuous ECIL. The output of thoracic duct lymphocytes decreases in a way that can be expressed as 2 exponential components. The first component represents a pool of readily accessible lymphocytes that are mobilized with a half time of about 11.2 days. The average size of this compartment in the 9 calves studied was estimated from the integral of the first 8 days to be 4.8×10^9 small lymphocytes/kg. This is about one half the size of the pool in rats reported by Caffrey *et al* (1962). The half depletion time of the less accessible lymphocyte pool cannot be determined with accuracy, but is of the order of 29.5 days.

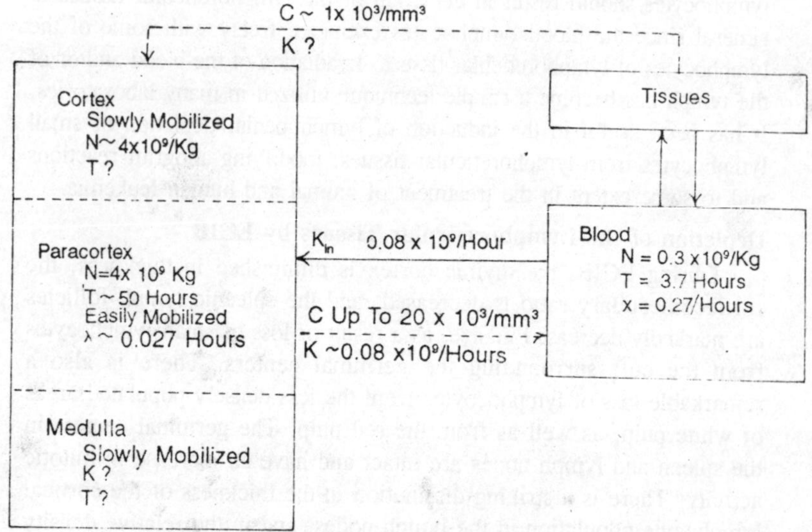

Fig. 4.6. Schematic presentation of the flow of lymphocytes from blood to tissues, to lymph nodes with estimates of time, fluxes and sizes where possible. N—number/kg body weight; T—average time in compartment; l—fractional turnover per hour; K—flux (number per hour per kg); C—concentration number per min.

During this period, about 1.2×10^9 lymphocytes/kg were killed. The number of lymphocytes estimated in both the readily accessible and the slowly mobilizable compartments is an overestimate, since we have no good means of determining the absolute production rate of cells in either compartment during the depletion period.

Studies on reestablishment of the pools following irradiation of the blood and lymph are incomplete at this time, but clearly follow a two-component process with an initial fast stage followed by a much slower stage. In some animals it has taken over a year for blood lymphocytes to attain pre-treatment levels. An analysis of the preceding data has been used to give a first approximation of the fluxes between blood and lymph nodes, the pool sizes and the time in pools. In blood the pool is about 0.3×10^9 cells/kg. The average time in the blood is 3.7 hours and about 27 per cent of the blood small lymphocytes are cleared per hour through the paracortical regions of the lymph nodes. In addition, an unknown flux of cells from blood into tissues is known to exist because of the return through the afferent lymphatics, but its magnitude cannot be measured. The pools of outer cortical lymphocytes and paracorticallymphocytes are both about the same size. The turnover rate of medullary and outer cortical lymphocytes is unknown.

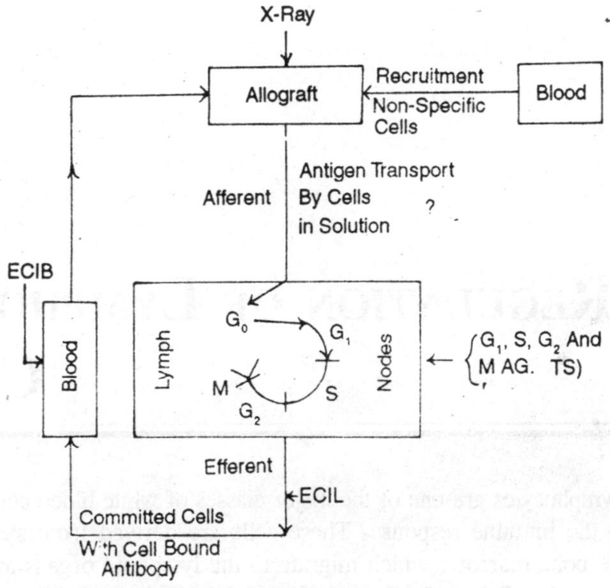

Fig. 4.7. Inhibition of immune response.

Although much has been learned about the migration pathways and proliferation of lymphocytes, little is known about the factors that regulate the system. The antigen is transported to the regional nodes either by cells or in solution. Within the nodes, there is transformation and proliferation and the chemotherapeutic agents active against the cell cycle can exert an immunosuppressive effect at this point. Activated cells migrate out in the efferent lymphatics where they can be attacked during ECIL or ECIB, thus reducing the flux of the activated cells through the allograft. At the allograft these cells can be killed by local irradiation and those that survive react with the antigen and kill graft cells either directly or indirectly. The inflammatory response results in recruitment of nonspecific cells into the area. The rapidity of the ultimate rejection of the graft is a function of the number and strength of antigenic differences between host and donor, the flux of activated cells into the graft, and the intensity of the inflammatory response set up by nonspecific factors that result in necrosis following thrombosis and in1pairment of circulation.

Hence, attack on the transplantation problem requires a knowledge of, and ability to control, lymphopoietic prolifertion and migration.

5

Regulation of Lymphoidal Cells

Lymphocytes are one of the major classes of white blood cells involved in the immune response. These cells are derived from stem cells in the bone marrow, which migrate to the lymphoid organs and develop into either B lymphocytes or one of the subclasses of T lymphocytes. Each type of lymphocyte plays a distinct and critical role in immunity. These functions are mediated in part by unique proteins with specific binding properties, called receptors, present on the surface of each type of lymphocyte. For example, B lymphocytes have cell surface immunoglobulin, or antibody, molecules which bind with a specific antigen. This binding stimulates activation and multiplication of the B lymphocyte. T lymphocytes also have cell surface, antigen-specific receptors which are distinct from immunoglobulin. Another example is the Fc receptor, found on certain lymphocytes, which binds the constant region of antibody molecules, called the Fc region.

If one examines a mixed population of lymphocytes microscopically, the cells are generally indistinguishable with respect to size, shape, etc. However, reagents have been developed which will identify the specific cell surface receptors so that they are visible by microscopy, and therefore allow visual identification of specific cell types. The two experiments described below demonstrate different reagents used to identify specific subpopulatlons of lymphocytes within a heterogenous population of lymphocytes, such as blood cells or spleen cells.

IDENTIFICATION OF Fc RECEPTOR-BEARING CELLS

Most mature B lymphocytes, as well as other accessory cells such as macrophages, possess cell surface receptors for the Fc portion of immunoglobulin, or antibody, molecules of the IgG class. Other cell types have receptors for the Fc region of other immunoglobulin classes. The Fc region of an antibody is also called the constant region, since all antibodies of a particular class have the same Fc region. In contrast, the Fab portion of an antibody contains the highly variable regions of the protein, where the specific antigen combining sites are formed.

The role of Fc receptors in the immune function of B lymphocytes remains unclear. On other immune cell types, however, Fc receptors may facilitate the destruction of cells or particles coated by antibodies during an immune response. It has been demonstrated that Fc receptors on macrophages promote binding and subsequent engulfing and destruction of antibody-antigen complexes which would damage host tissues if allowed to accumulate. Another important example is the Fc receptors for IgE antibodies found on basophils and mast cells, which are probably involved in allergic reactions.

Cells bearing Fc receptors on their surface can be identified easily by the formation of rosettes with antibody-coated erythrocytes (red blood cells). This procedure involves mixing a suspension of lymphocytes, for example spleen cells, with erythrocytes which have been coated with IgG antibodies produced against the erythrocytes. The Fab portion of the antibodies will attach to the erythrocyte surface. The Fc portion of the antibodies will be exposed, and will bind with the IgG Fc receptors on the B lymphocytes and other cells (Figure 2). Cells bearing Fc receptors are then identified microscopically by the cluster of erythrocytes around them, called a rosette. Rosetted cells can be separated from cells which have not formed rosettes by differential centrifugation techniques. Removal of the erythrocytes then yields a cell population enriched for Fc receptor-positive cells. Variations on this procedure are commonly used to isolate human lymphocyte populations inn clinical labs, and to isolate or deplete various white blood cell populations in research labs.

In this module, sheep erythrocytes, which have been coated with IgG antibodies produced against sheep erythrocyte stroma proteins, will be used to identify IgG Fc receptor-bearing cells in a population of cells isolated from mouse spleen. The mouse Fc receptors for IgG will bind with the Fc portions of either mouse or rabbit IgG molecules.

The percentage of mouse spleen cells which have Fc receptors for IgG on their surface can then be calculated.

Safety Guidelines

Standard lab safety practices should be followed.

Experimental Outline

Timetable of events

The entire lab procedure, from the mixing of lymphocytes and coated erythrocytes to cell counting, can be completed in less that one hour. This procedure may be performed in the same two to three-hour lab period as another short experiment. Alternatively, students, rather than the instructor, may perform the antibody coating of erythrocytes, or any other preparation steps, during the lab period. Spleen cells should be harvested and erythrocytes removed on the day of the lab. The sheep erythrocytes should be prepared within one week prior to the lab and stored in the refrigerator.

Pre-lab Preparation

Spleen cell preparation:	30 minutes
Removal of erythrocytes from spleen cell (hypotonic or acetic acid lysis):	30 minutes
Preparation of erythrocytes:	40 minute (55 min. if making HBSS from scratch)

Antibody coating of erythrocytes: 55 minutes

Lab Procedure

Generate rosettes: Mix spleen cells and antibody-coated erythrocytes and incubate; stain with crystal violet - about 25 minutes.

Observe and quantitate rosettes - about 15 minutes.

Materials

NOTE: If the use of mice for this lab is not appropriate, it is suggested that Experiment B be performed instead of this one, since either bovine spleen (from a slaughterhouse) or cultured lymphocytic cell lines can be used. It may be possible to use such cells in this experiment, but that has not been tested yet by the author.

Antiserum produced in rabbits or mice against sheep erythrocytes (1 ml)

Sheep blood or glutaraldehyde-fixed sheep erythrocytes (4 ml)

Hank's Balanced Salt Solution (HBSS) (250 ml)

Lymphocytes isolated from mouse spleen, purified by acetic acid lysis to remove erythrocytes, 1 x 10^6 cells/ml (See below for alternative protocol for removing erythrocytes: hypotonic lysis)

5% suspension, in Hanks balanced salt solution (HBSS), of sheep erythrocytes coated with IgG antibodies

Crystal violet, 0.5% w/v in distilled water

17 × 100 mm clear tubes, either with caps or with parafilm for sealing

Microscope slides and cover slips

5 ml pipets

1 ml pipets (or automatic micropipet)

Pipet pumps

37°C water bath

12 × 75 mm glass tubes

Microscope capable at least 200x total magnification

Pasteur pipets and bulbs

Sources

Sigma Chemical Co.

P.O. Box 14508

St. Louis. MO 63178; 1-800-325-3010

For indirect, fluorescent labeling:

#M7019 Goat antibody to mouse IgM

#F9259 FITC-Iabeled rabbit antibody to goat immunization

For indirect, enzyme labeling:

#M7019 Goat antibody to mouse IgM

#A7650 Alkaline phosphatase-labeled rabbit antibody to goat immunoglobulin

For direct, fluorescent labeling:

#F9259 FITC-Iabeled goat antibody to mouse IgM

Also, antibodies against IgM from other species are available if cells other than mouse are used.

American Type Culture Collection (ATCC)

12301 Parklawn Drive

Rockville, MD 20852; 1-800-638-6597

Possible cultured mouse cell lines-in place of spleen cells:

Surface IgM positive:

#ATCC 1702, WEHI-231

#ATCC CRL 1704, WEHI-279

#ATCC TIB 209, x16C8.5

(All are mouse B cell lymphomas.)

Surface IgM negative:

#ATCC TIB 47, BW5147.3

#ATCC TIB 155, LBRM-33

(Both T cell lymphomas.)

Pre-lab Preparation

1. Preparation of Hank's Balanced Salt Solution: This can be purchased either as a 1 x or 10x solution or as a powder. If prepared in the laboratory, the following recipe will make one liter of 1x HBSS:

   Solution A: 1.0 g D-glucose
   .01 g Phenol red
   .06 g KH_2PO_4
   .09 g $Na_2HPO_4 \cdot 7H_2O$

   Solution B: .14 9 $CaCI_2$
   0.4 9 KCI
   8.0 g NaCI
   0.1 g $MgCI_2 \cdot 6H_2O$
   0.1 g $MgSO_2 \cdot 7H_2O$
   0.35 g Sodium bicarbonate

   Separately dissolve the components for Solution A and Solution B, each in 300 ml distilled water. Mix Solution A and Solution B together. Make sure all ingredients are in solution, then add Sodium bicarbonate. Check to make sure that the pH is approximately 7.4, adjusting if necessary. Store in refrigerator.

2. Preparation of sheep erythrocytes:
   (a) In a conical 50 ml centrifuge tube, mix 4 ml sheep blood or reconstituted erythrocytes and 16 ml HBSS.
   NOTE: If a centrifuge capable of spinning 50 ml tubes is not available, divide this into two 15 ml tubes.
   (b) Centrifuge for 10 minutes, at 400 x g (usually setting 4 on a table-top clinical centrifuge).
   (c) Discard supernatant and resuspend cells in 20 ml HBSS.
   (d) Repeat steps b and c twice.
   (e) Store washed erythrocytes in refrigerator for no more than one week.

3. Antibody coating of erythrocytes:
   (a) Dilute the antiserum to sheep erythrocytes by mixing 1 ml antiserum with 19 ml HBSS.
   (b) Mix antiserum with washed sheep erythrocytes in conical centrifuge tube(s).
   (c) Incubate mixture 30 minutes at 37°C.

       NOTE: The timing and temperature of this incubation are not critical.
   (d) Centrifuge mixture 10 minutes at 400 x g.
   (e) Discard supernatant and resuspend cells in 40 ml HBSS.
   (f) Repeat steps d and e once.
   (g) Store coated erythrocytes in refrigerator for one week.
4. Hypotonic lysis to remove erythrocytes from spleen cells (alternate protocol):
   (a) Centrifuge spleen cells (isolated as in Unit 6 Module 2) 10 minutes, 400 x g.
   (b) Decant supernatant.
   (c) Tap cell pellet to resuspend cells in small amount of supernatant remaining.
   (d) Add 9 ml distilled water to cell pellet. After cells have been in water for 2 seconds, quickly add 1 ml of 10x concentrated HBSS or PBS and mix thoroughly with a pipet.

       NOTE: The timing of this step is important. The lymphocytes are somewhat more resistant to the hypotonic conditions than the erythrocytes, but the lymphocytes will also be lost if exposed to the water for too long.
   (e) Wash the lymphocytes by cehtrifuging as in step a., resuspending the pellet in 1 x HBSS and centrifuging again. The cell pellet should be white rather than red at this point. If there is still a lot of red in the pellet, repeat steps a-e to remove additional erythrocytes.
   (f) Resuspend the cells in 1 x HBSS and count.

Method

1. Mix 2.5 ml of lymphocyte suspension with 2.5 ml antibodycoated erythrocytes in a 17 x 10 mm tube.
2. Seal the tube tightly with a cap or with parafilm. It is very important that you are able to invert the tube without any liquid spilling out. Even if your tube has a cap, you may want to wrap parafilm around the cap to seal it.

3. Warm the tube to 37°C by placing it in the water bath for 5 minutes.
4. Remove the tube from the water bath and gently mix it for 10 minutes by inverting it back and forth. Keep your hand around the tube during mixing to keep it warm.
5. Place the tube on ice until ready to count the cells.
6. For counting the cells, stain the white blood cells with crystal violet for easier visualization as follows:
   (a) Using 1 ml pipet or automatic micropipet, remove 0.1 ml from the tube containing lymphocytes and coated erythrocytes and transfer to a 12 x 75 mm tube.
   (b) Dilute sample with 0.4 ml HBSS.
   (c) Add 1 drop 0.5% crystal violet to sample using pasteur pipet.
   (d) Let the tube sit 5 minutes at room temperature to allow the lymphocytes to take up the stain.
7. Place a drop of the stained cells on a microscope slide and place a cover slip on top of the drop.
8. Place the slide on a microscope and focus on the cells, first using the 10x objective, and then using the 20x objective. You should be able to see both naked, stained cells, and white blood cells which are surrounded by a cluster of pale, smaller cells. These clusters are the rosettes.
9. Count the number of rosettes visible in the field. Also count the total number of white blood cells visible. Record these two numbers. You should count a large enough sample to obtain a total of at least 50 white blood cells. If necessary, count more than one field and add the numbers obtained.

   NOTE: Using the 10x objective allows you to count more cells in a single field, but it may be easier to see the cells using the 20x objective. Choose whichever objective you find easiest to use to carry out the cell counting described in step 9.

Results

1. Draw diagrams illustrating how the two types of white blood cells, rosetted and non-rosetted, appear under the microscope.
2. Use the cell counts obtained in step 9 to calculate the percent of lymphocytes in the spleen cell population which are Fc receptor positive.
3. If the concentration of erythrocytes in the mixture is too high, they may obscure visualization of the lymphocytes and rosettes. If this seems to be a problem, dilute the sample further in HBSS.

4. Washing the erythrocytes after antibody coating is an important step, since residual excess antibody will inhibit rosette formation.

Identification of B Lymphocytes by Detection of Cell Surface Immunoglobulin

The B lymphocytes are the cells that produce specific antibodies when activated during an immune response. While antibodies are secreted during an active immune response, most B lymphocytes contain antibody (immunoglobulin) which remains attached to the B lymphocyte surface. This immunoglobulin, which is of the IgM class in inactivated B cells, is produced when the B cells develop in the bone marrow. Before the IgM protein is produced, the genes encoding the IgM heavy and light chains undergo rearrangement to generate unique variable regions which will combine with a specific antigen. Since this rearrangement occurs differently in each developing B cell, each B cell will produce IgM antibodies with a unique antigen specificity. Binding of the specific antigen with the IgM on a particular B cell will stimulate that cell to become activated and to begin dividing as part of the immune response to the antigen.

The presence of surface IgM can be used to identify B lymphocytes. As with many cell surface proteins, specific antibodies against the IgM class of immunoglobulin can be used as a tool to mark IgM-bearing cells. These antibodies are generated by injecting purified IgM proteins into a different species of animal and isolating the antibodies from the animal's serum. Usually antibodies which react with constant regions of IgM (i.e. determinants that are found on all IgM molecules, regardless of antigen specificity) are selected. The antibodies produced in this manner will bind to the IgM present on most B lymphocyte surfaces. If these antibodies are labeled, such as with a fluorescent dye or with an enzyme that will produce a colour when it acts on an appropriate substrate, the B lymphocytes can be distinguished microscopically from the rest of the cells in a population. The IgM antibodies may be labeled directly or, alternatively, a stronger signal can be achieved by using unlabeled antibody against the IgM (called the primary antibody) in the first step followed by labeled antibody produced against the primary antibodies (called the secondary antibody) in a second step. This indirect labeling method is used frequently in clinical and research laboratories to detect the presence of many types of cell surface proteins on subpopulations of cells. This detection system can be used to determine the percent of cells bearing the cell surface protein, as well as to isolate or deplete that cell population.

Instructions are provided for both fluorescence and enzyme labeled second antibody. The instructor will indicate which label will be used. Be sure to follow the appropriate directions for the label being used. The use of a fluorescent label requires the availability of a fluorescence microscope, while the enzyme labeled antibody can be visualized with a standard light microscope. Note that this procedure is written for detecting IgM on B lymphocytes isolated from mouse spleen. However, the instructor may choose to use cells isolated from a different animal, or cultured cells. If so, additional instructions and corrections to the protocol will be provided.

Safety and Waste Disposal

1. General laboratory safety guidelines should be followed throughout the procedure.
2. If using the fluorescence microscope, the eyes should be protected from the ultraviolet light used to excite the fluorescent dye. Most fluorescent microscopes are equipped with protective shields that are positioned so that the eyes of the person looking into the objectives are protected.

Experimental Outline

Timetable of events

The lab protocol for indirect labeling, through cell observation and counting, can be completed in a 3-hour lab period. It may be possible to complete this in a 2-hour lab period if there are no unscheduled waiting times, etc.

Pre-lab Prep

Preparation of PBS-BSA	10 min. (25 min if making PBS)
Preparation of substrate	10 min.
Washing and counting cells	20 min. (60-75 min if starting with spleens)
Dilution of antibodies	15 min. (2-3 hours if testing antibody dilutions)

Lab Procedure

Primary antibody treatment. Mix cells and antibody against IgM and wash away unbound antibody - about 45 min.

Secondary antibody treatment. Mix primary antibody treated cells with labeled secondary antibody, wash away unbound secondary antibody - about 45 min.

Observation and counting of labeled cells. Calculate percent of IgM bearing cells - about 15 min.

Materials

1. If the use of mice is not appropriate for this lab, there are alternative sources of lymphocytes. Spleen from several different species can also be used, since antibodies against IgM are available (from Sigma Chemical Co.; see Resources). If a nearby research lab will be using other parts of an animal you might request the spleen. Bovine spleen may be available from a local slaughterhouse. Alternatively, cultured lymphocytic cell lines are available from ATCC or from research labs. A mixture of cultured IgM-positive, B lymphocytic cells and IgM-negative cells (such as a T lymphocytic cell line,) mixed in approximately 50:50 proportions, would produce results similar to splenic lymphocytes. Wash cultured cells twice with HBSS, count, and adjust concentration to 1 x 10^6/ml. The protocol described here assumes the use of mouse spleen cells, but the procedures would be the same for other species; just substitute antibodies with appropriate specificities.
2. If a fluorescent microscope is available, the fluorescence detection method is a useful lab technique for students to experience, and it generally yields stronger detection signals than the enzyme label. However, unless multiple fluorescence microscopes are available, students will have to wait in line to observe their stained cells. The stained cells can be stored, covered with foil, in the refrigerator for a few days for later observation.
3. A direct protocol may be used for fluorescence detection, using FITC-labeled rabbit antibody to mouse IgM. This will eliminate one incubation and washing step. However, the fluorescence obtained is generally weaker and more difficult to observe. Direct staining is also possible with enzyme-labeled antibodies against mouse IgM, but this may not yield detectable signals.
4. The success of this protocol depends critically on using the appropriate dilution of both primary and secondary antibodies. The manufacturer generally recommends a working dilution. However, it is best to determine the optimum dilution empirically for each antibody. This can be worked into the lab exercise, by having each group use a different dilution, and having all groups examine the results from the most successful group(s). Alternatively, if enough supplies are available each group can prepare multiple samples using a range of dilutions encompassing the recommended dilution. Phosphate-buffered saline with 0.1 % bovine serum albumin (PBS - BSA) 100 ml

Lymphocytes isolated from mouse spleen and purified by acetic acid lysis or hypotonic lysis, 2×10^6/ml in PBS-BSA (2×10^7 cells total)

12×75 mm centrifuge tubes, 10

Microscope slides, 20-30

Cover slips, 20-30

Pasteur pipets, 30-40

Bulbs, 10

Goat antibodies against mouse IgM, diluted appropriately in PBS - BSA (see Note #4 above), 2 ml

1 ml pipets, 30-40

Pipet pumps, 10

Gloves

Ice

90% glycerol in PBS-BSA

For fluorescence detection

Fluorescein isothiocyanate (FITC)-Iabeled rabbit antibodies against goat immunoglobulin, diluted appropriately in PBS-BSA, 2 ml

For enzyme-linked colour detection

Alkaline phosphatase-labeled rabbit antibodies to goat immunoglobulin, diluted appropriately in PBS-BSA, 2 ml

Alkaline phosphatase substrate (nitroblue tetrazolium (NBT) and 5-bromo-4-chloro-3-indolylphosphate (BCIP))

For fluorescent detection

Fluorescein isothiocyanate (FITC) labeled goat antibody against rabbit immunoglobulin (secondary antibody)

For enzyme linked colour detection

Alkaline phosphatase-labeled goat antibody against rabbit immunoglobulin (secondary antibody)

Alkaline phosphatase substrate (nitroblue tetrazolium (NBT) and 5-bromo-4-chloro-3-indolylphosphate (BCIP)), made fresh by adding 6.5 μI BCIP and 33 μI NBT to 5 ml PBS-BSA

Pre-lab Preparation

1. *Preparation of PBS-BSA*: To 100 ml PBS, add 0.1 g bovine serum albumin, stir to mix.

   PBS can be purchased as a 1x or 10x solution or as a powder. If prepared in the laboratory, dissolve the following in 150 ml distilled water:

2.19 g NaCI

1.28 g KH_2PO_4

2.63 g $Na_2HPO_4 \cdot 7H_2O$

Adjust pH to 7.2, add distilled water to bring volume to 250 ml.

2. *Hypotonic lysis to remove erythrocytes from spleen cells*:
   (a) Centrifuge spleen cells (isolated as in Unit 6 Module 2) 10 minutes, 400 x g.
   (b) Decant supernatant.
   (c) Tap cell pellet to resuspend cells in small amount of supernatant remaining.
   (d) Have ready a pipet or syringe containing 1 ml of 10x concentrated HBSS or PBS.
   (e) Add 9 ml of distilled water to the cell pellet. After cells have been in water for 2 seconds, quickly add 1 ml of 10x concentrated HBSS or PBS and mix thoroughly with a pipet. NOTE: The timing of this step is important. The lymphocytes are somewhat more resistant to the hypotonic conditions than the erythrocytes, but the lymphocytes will also be lost if exposed to the water for too long.
   (f) Wash the lymphocytes by centrifuging as in step a., resuspending the pellet in PBS-BSA, and centrifuging again. The cell pellet should be white rather than red at this point. If there is still a lot of red in the pellet, repeat steps a-f to remove additional erythrocytes.
   (g) Resuspend the cells in PBS-BSA and count. Adjust the cell concentration to 2 x 10^6/ml.
3. *Dilution of primary and secondary antibodies*: If desired, the instructor can prepare a series of dilutions and test them prior to the lab, and use the optimal dilution in the lab. Antibodies should be diluted fresh, no more than 12 hours before the lab. If desired, aliquot diluted antibodies into labeled tubes for groups. FITC-Iabeled antibodies should be protected from light as much as possible.
4. *Preparation of cells*: Cells should be washed once in PBS-BSA by centrifuging 10 minutes at 250 x g, resuspending in PBS-BSA, and centrifuging again. Resuspend cell pellet in PBS-BSA so that cells are at a concentration of 2 x 10^6/ml.

5. *For alkaline phosphatase label-preparation of substrate*: Both substrate components are available as tablets from Sigma Chemical Co., which are dissolved in the indicated amount of distilled water just before use. Make substrate solutions as fresh as possible, and protect from light until used.

Method

NOTE: It is important to keep the cells and reagent on ice throughout the procedure. If the cells become warm, the IgM-antibody complexes may be internalized and will be difficult to visualize. The addition of 0.2% sodium azide to the PBS-BSA would inhibit internalization, but since azide is highly toxic it is not recommended for use in a classroom laboratory.

1. Transfer 1 ml of lymphocyte suspension to a 12 x 75 mm centrifuge tube.
2. Centrifuge 10 minutes at 250 x g.
3. Carefully remove supernatant with pasteur pipet and discard.
4. Add 0.1 ml diluted antibody against IgM and gently resuspend the cell pellet by tapping the bottom of the tube or by using the pipet to break up the pellet.
5. Incubate on ice 15 minutes.
6. Add 1 ml PBS-BSA. Mix.
7. Centrifuge 10 minutes, 250 x g
8. Remove supernatant and discard.
9. Add 1 ml PBS-BSA and resuspend the cell pellet.
10. Repeat steps 7 and 8.
11. For fluorescent detection (see below for enzyme linked colour detection) :
    (a) Add 0.1 ml FITC-labeled goat antibody against rabbit immunoglobulin and resuspend the cell pellet.
    (b) Incubate 20 minutes on ice.
    (c) Wash cells as in steps 6-10 above.
    (d) Add 0.1 ml PBS-BSA and resuspend the cell pellet.

    For enzyme linked colour detection:
    (a) Add 0.1 ml goat antibody against rabbit immunoglobulin and resuspend the cell pellet.
    (b) Incubate 20 minutes on ice.
    (c) Wash cells as in steps 6-10 above.
    (d) Add 0.1 ml peroxidase substrate and resuspend the cell pellet.

12. Place a drop of 90% glycerol on a microscope slide. Add a drop of stained cell suspension. Place a cover slip over the cells, taking care to avoid air bubbles.
13. For fluorescence detection:
    (a) Observe cells using microscope equipped with ultraviolet light source and filters for fluorescence detection. Use a 20x or 40x objective.
    (b) First count the total number of lymphocytes using normal light. The lymphocytes will be small, round cells with little cytoplasm.
    (c) Then, in the same field, use the ultraviolet light to count the number of fluorescent cells. The fluorescence should have a "patchy" appearance on the lymphocyte surface.

    For enzyme linked colour detection:
    (a) Observe cells using a 20x or 40x objective.
    (b) Count the total number of lymphocytes in the field. The lymphocytes are small, round cells with little cytoplasm.
    (c) Count the total number of lymphocytes which have stained with peroxidase. These cells will have a purple stain over the surface.

Results

1. Use the cell counts obtained to calculate the percent of B lymphocytes In the lymphocyte sample.
2. Is the percent of B lymphocytes calculated from your data consistent for what one would expect to see in this cell sample? Explain.
3. The critical parameter is antibody dilution. No positive cells may indicate that one of the antibodies is too dilute. A high degree of background staining (resulting in no distinction between positive and negative cells) may indicate the antibodies are too concentrated and nonspecific binding is occurring.
4. You may want to include a control sample which is treated with the secondary antibody, but not the primary antibody. This may help determine the degree of nonspecific binding of the secondary antibody.
5. Another way to reduce nonspecific staining is to add one more washing step between each antibody addition. This will increase the time required for the exercise.

Identification of Lymphoid Cells in Blood Smears and Tissue Sections

Blood cells undergo progressive differentiation from stem cells originating in the bone marrow. In the presence of differentiation inducing growth factors B lymphocytes mature in the bone marrow and T lymphocytes mature in the thymus; both of these lymphoid organs are classified as primary lymphoid organs. Peripheral blood, lymph nodes, the spleen, and the tonsils are classified as secondary lymphoid organs since they contain mature, fully functional lymphocytes. By comparing cells from the bone marrow with cells from the peripheral blood and the lymph nodes some of the cytological differences between immature stem cells and mature lymphocytes may be observed.

The bone marrow contains lymphoid and myeloid cells at various stages of differentiation. The larger more primitive cell types show less dense nuclei and are more closely related to immature stem cells. These undifferentiated cells are nonfunctional and differences between lymphoid and various myeloid cell types are more difficult to distinguish.

Peripheral blood smears show well defined different cell types. Functional myeloid cells include erythrocytes, granulocytes, monocytes and platelets. Lymphocytes show a circular dark staining nucleus under normal conditions. It is not possible to distinguish T lymphocytes from B lymphocytes in peripheral blood using standard Wright's-Giemsa stain.

Distinct circular regions of cells in the lymph nodes called germinal centers are present near the outer, cortical edge of the organ. These centers represent sites where antigen processing cells like macrophages have stimulated B lymphocytes with antigen and induced clonal expansion of those B lymphocytes that were able to respond to antigen. T lymphocytes, including key T helper cells, are located in the lymph node in the paracortical region that lies closer to the central portion of the lymph node.

There may be a need to use standard definitions for the histology and cellular terms. A proposed set of definitions follows:

B-cell. Lymphocyte that is the key cell in the production of antibodies following stimulation

Bone Marrow. Source of stem cells for all cellular components of blood.

Clonal Expansion. The stimulation of a single cell to undergo cell division and produce clonal descendants

Cortical. Outer layer or region of an organ or tissue

Differentiation. The process of selective gene activation as a cell progresses in development from a stem cell

Erythrocyte. Differentiated red blood cells

Germinal Centers. Areas of lymphoid tissue where stimulated lymphocytes are dividing

Granulocytes. Phagocytic white cells that show prominent cytoplasmic granules and are abundant in peripheral blood

Lymph Nodes. Secondary lymphoid organs along length of lymphoid vessels that act as sites for antigen trapping and B cell stimulation

Macrophage. A white cell derived from a monocyte that helps process and present antigen

Monocyte. A white cell type found in peripheral blood that, like more abundant granulocytes, has phagocytic activity

Myeloid Cells. A family of cells which develop from a bone marrow stem cell to become erythrocytes, monocytes, granulocytes, and platelets

Paracortical Region. A region outside of the cortex of an organ or tissue, often farther from the external edge

Spleen: A secondary lymphoid organ which stores white cells and red blood cells

Thymus. A primary lymphoid organ which is essential for the differentiation of T lymphocytes

Tonsils. Secondary lymphoid organ that stores differentiated lymphocytes

T Lymphocyte. Lymphocyte that is a key cell in stimulating immune cells and forming T killer cells for rejecting grafts and tumors

T helper cell. Type of T lymphocyte that helps stimulate both functional T and B cells through production of soluble effectors called *lymphokines*

Safety Guidelines

1. Use commercial slides and do not make peripheral blood smears using finger sticks.
2. Clean the oil immersion lens after use by using only microscope grade lens tissue.
3. Before trying to remove slides from the microscope stage rotate back to low power lens to provide clearance.

Experimental Outline

1. Identify the different mature cell types in peripheral blood.
2. Identify different immature cell types in the bone marrow and compare these to peripheral blood.

3. Identify the overall structure of the lymph node and the location and morphology of lymphoid cells in the organ. Distinguish T from B lymphocytes based on their location.

Materials

Commercial slides of peripheral blood, Wright's-Giemsa stained.

Commercial slides of lymph nodes, Hematoxylin-Eosin stained.

Commercial slides of bone marrow smears, Wright's stain, or white bone marrow, Hematoxylin-Eosin stained.

Microscopes with oil immersion lenses

Immersion oil

Microscope lens tissue

Pre-lab Preparation

1. Obtain slide sets of peripheral blood, bone marrow, and lymph nodes, microscopes with oil immersion lenses, immersion oil, and microscope lens cleaning tissue.
2. During pre-lab discussion draw characteristic lymphoid cells and general architecture of the tissues.
3. Instruct students in the proper sequence of steps in using the oil immersion lens.

Method

1. For peripheral blood smears move the slide under high power to the feather edge end of the smear where the lower density of cells does not lead to clumping. Switch to oil immersion and make minor fine focus changes after contact of the lens with the oil drop on the slide. Draw the cell types visible.
2. Follow a similar procedure for bone marrow smears and draw the cell types observed.
3. For tissue sections get an overall orientation of the tissue regions at low power magnification. Proceed up through high power and then to oil immersion lenses. Draw the cell types present. For lymph nodes draw cells seen in the germinal centers and in the paracortical region.

Results

Compare the drawings you have made from the three different lymphoid organs.

6

Diversity in Lymphocyte Receptors

B and T lymphocytes are two types of leukocytes (i.e. white cells) that are responsible for antigen specific immunity. B cells express antibody molecules on their surface as antigen receptors, and will develop into antibody secreting plasma cells upon activation by their corresponding antigens. T cells also carry surface receptors for antigens, termed simply T cell receptors or TcR. They include helper T cells, which 'help' B cells make antibodies, and cytolytic T cells which kill target cells specifically.

An antibody (i.e. immunoglobulin or Ig) is a protein molecule with two identical heavy (H) chains and two identical light (L) chains. Both the Hand L chains consist of a variable (V) region that constitutes the antigen combining site, and a constant (C) region that determines the isotype (e.g. IgG, IgM, IgA, IgD and IgE for H chains; k, λ for L chains) of an antibody. T cell receptors (TcR) have a heterodimeric protein structure which exists either as a α/β or a γ/δ form. Each (TcR) polypeptide subunit (i.e., α, β, γ, δ) also carries an antigen-recognizing V region, and a C region of unknown function.

It is generally assumed that a human body (other mammalian species as well) is able to produce enough antibodies or TcR, each with a different specificity, to cope with all foreign antigens that may ever exist (possibly in the range of 10^8). This assumption, however, does not agree with the estimate that the human genome has only approximately 10^6genes in reserve. How, then, can our immune system possibly generate such a diversity that far exceeds our total genetic capability? An answer to this longtime puzzle has become clear only recently, due largely to the molecular biology studies pioneered by Susumu Tonegawa and others. It was discovered that a considerable diversity of T and B cell specificities could be generated by recombining scattered, discrete segments (i.e. DNA sequences that represent "minigenes") of Ig genes or TcR genes. The gene rearrangement events occur in precursor cells during their development into mature T or B cells. Once completely rearranged, the DNA sequence (i.e. specificity) of a full-length Ig gene (of a mature B cell) or a TcR gene (of a mature T cell) is permanently fixed.

Take for example the generation of a complete gene that will code for a full length heavy (H) chain of an antibody. A complete H gene consists of a variable (V) region gene segment, a diversity (D) gene segment, a joining (J) gene segment and a constant (C) region gene segment. The juxtaposed (V-D-J) gene segments jointly code for the variable region of the H chain, which determines antibody specificity; the C gene codes for the constant region which does not contribute to antibody specificity. Before H chain gene rearrangement there are by estimate approximately 250 V, 10 D and 5 J gene segments in each precursor cell of a B lymphocyte, During development into a B cell one V, one D and one J gene segment will combine In a precursor cell, and there are potentially 12,500 (i.e. 250 × 10 × 5) ways a combination could result! (It works in a way like the number-locks.) Similarly, by combining at random the V and J gene segments (approximately 200 V and 10 J) of the light chain (L) genes, B cells can generate 2,000 L chain combinations. Pairing of Hand L chains (H x L) could thus provide up to more than 10^7 specificities. Additional diversity is created by imprecise joining of the various gene segments, and by point mutation, altogether leading to billions of antibody possibilities. Similarly, during T cell development TcR gene segments in precursor cells undergo rearrangement, resulting in the generation of enormous diversity in T cell specificities.

In this module, you will perform a 'dry lab' to investigate the gene rearrangement mechanism by which the immune system generates diversity in antigen recognition by B cells (i.e. using antibody) and T cells (i.e. using TcR). Paper clips with different colours will be used to represent different minigenes or DNA segments of Ig and TcR genes. You will be instructed to combine these paper clip genes to construct a series of complete IgH genes and TcR-β genes that code respectively for antibodies and T cell receptors of different specificities.

Safety Guidelines

None, except avoid sticking your fingers with the sharp ends of paper clips,

Experimental Outline

Each of the two exercises could be completed within an hour. It is recommended that Exercise A, and Exercise B be performed in sequence.

Sort out coloured paper clips

Perform Experiment A: Ig gene rearrangement

Perform Experiment B: TcR gene rearrangement

Group paper clips according to colour and return to instructor

Materials

Paper Clips: red (R), pink (Pi), orange (O), yellow (Y), blue (Bl), green (G), purple (Pu), white (W), black (B), silver (S). Each paper clip represents a "mini-gene" or gene segment.

(Alternatively, paper clips tagged with colour adhesive tapes, or strings or beads with corresponding colours can be used.)

Pre-lab Preparations

No special pre-lab preparations are required.

Method

Experiment A

Immunoglobulin (Ig) gene rearrangement

1. Obtain paper clips with different colours from instructor.
2. Use the following letter keys for the paper clips.

Black – B	Blue – BI	Green – G
Orange – O	Pink – Pi	Purple – Pu
Red – R	Silver – S	White - W
Yellow – Y		

3. Assume that each paper clip represents a gene segment, use your paper clips to represent discrete DNA segments of Ig heavy chain (H) genes as follows:

   Variable region (V) genes: V_1 (Bl), V_2 (G), V_3 (O), V_4 (Pi)

   Diversity region (D) genes: D_1 (Pu), D_2 (R)

   Joining region (J) genes: J_1 (W), J_2 (Y)

   Constant region (C) gene: C (B)

   Use the silver (s) paper clips to represent intervening sequence (i.e. introns).

4. Arrange your paper clips to construct a hypothetical embryonic or germ line DNA sequence of Ig gene before rearrangement.
5. Each complete heavy chain gene consists of one V gene, one D gene, one J gene and one C gene. Using your fingers as recombinase enzymes, construct a complete heavy chain gene by combining one of the four V genes, one of the D genes, one of the J genes and the C gene (for simplicity, only one C gene is considered in this exercise). You can assume that any V gene can combine with any D gene, which in turn can combine with any J gene. Record the recombinant gene sequence (e.g., V_1-D_2-J_3-C). Return the paper clips to the original DNA sequence.
6. Again, follow the instruction in (5) and construct another complete heavy chain gene sequence with a configuration different from the previous one. Record the new sequence. Return the paper clips to the original DNA strand.
7. Assume that you have been infected by five different infectious agents and your immune system responded by producing a different Ig heavy (H) chain specific for each invader. Demonstrate how this could be achieved by combining the 'paper clip' genes to construct five complete H genes that will code for 5 distinct heavy chains (representing five specificities). Record your five combinations as in (4).
8. Disconnect your paper clip chains and group the paper clips by colour.
9. Proceed to EXPERIMENT B.

Experiment B

T cell receptor (TcR) gene rearrangement

1. As in EXPERIMENT A
2. Assume that each paper clip represents a gene segment, use your paper clips to represent discrete DNA segments of TcR β chain genes as follows:
   Variable regions (V) genes: V_1 (B), V_2(W), V_3(Y), V_4(Pu)
   Diversity region (D) genes: D_1 (R), D_2(G)
   Joining region (J) genes: J_1 (O), J_2(Pi)
   Constant region (C) genes: C(BI)
   Intervening sequence: Intron (s)
3. Arrange your paper clips to assume a hypothetical germ-line configuration of TcR β gene sequence before rearrangement.

4. As assumed in Experiment A. you have been infected by five different infectious agents and your B cells responded by producing an antibody specific for each organism. Again, assume that for antibody production each antibody producing B cell must collaborate with a T helper cell carrying a surface receptor (i.e., TcR) specific for the corresponding organism. Using the paper clip gene sequence, construct five complete genes that could code for five distinct TcR β chains. As with the Ig heavy chain, the complete gene for a TcR β chain includes one V, one D, one J and one C region gene. Record the five combinations you constructed.

Results

1. Using a format as shown in the example below, illustrate in parallel the DNA sequences of all the five complete Ig heavy chain genes you constructed in step 4 of the Experiment A. Try using colours that correspond to the gene segments to accentuate the homologous and non-homologous regions among the five different DNA sequences.

   V_1 D_1 J_1 C
2. As instructed above, illustrate in parallel the DNA sequences of all five complete TcR β chain genes you have constructed in step 4 of the Experiment B.
3. Do you think your results demonstrate gene rearrangement as an effective mechanism to generate diverse gene sequences? Discuss your opinion(s).
4. As a conclusion of your experiments, write a paragraph to compare the strategies employed by T and B cells to generate diversity in their receptors for antigens.

7

Enzyme-linked Immunosorbent Assay

The body's immune system plays an important role in host defense mechanisms against foreign invaders. One major response of the immune system to infection is the production of antibodies, (i.e., *immunoglobulins* or Ig) in response to the stimulation of antigen(s) of the infecting agent. The antibody formed reacts specifically to the antigen, leading to the eradication of the particular infecting agent and no other. The antibody proteins circulate in the blood stream and can be detected in the serum of the infected individual.

Many laboratory procedures (i.e., assays) have been developed to measure antibodies and to detect the specific interaction between antibodies and antigens. These immunoassays include immuno-precipitation test (e.g. gel diffusion), agglutination test, complement fixations test, immunofluorescence test, *radioimmunoassay* (RIA) and enzyme-linked immunosorbent assay (ELISA), Western Immunoblot assay, etc. Using these immunoassays the antibody response of a host against infection can be followed, and the results obtained can contribute to the diagnosis of the infection. For example, detection of serum antibody to HIV (AIDS virus) is considered a reliable criterion to identify infected individuals who have become seropositive.

Although initially developed for antibody measurement, immuno-assays have also been adapted successfully to detect and quantitate antigen in unknown samples. For example, antibodies specific for HIV antigens can be produced by injecting the antigens into laboratory animals (e.g. rabbits), and the specific antibodies obtained can then be used to develop immunoassays for measuring HIV antigens in infected individuals. Due to their high specificity and sensitivity immunoassays are commonly used for a great variety of measurements of both antibodies and antigenic analytes (e.g. hormones, drugs, etc.) in research, analytical and diagnostic laboratories.

The basis of all immunoassays is an interaction between an antibody and a corresponding antigen. The interaction can in turn be conveniently detected by conjugating a measurable label to either the antibody or the antigen, such as a radioisotope, a fluorescent compound or an enzyme. An enzyme label is generally favoured over either fluorescent compounds which require expensive equipment for detection, or radioisotopes which are a biohazard and unstable. Thus, in recent years, the enzyme-based immunoassay, generally referred to as enzyme linked immunosorbent assay (ELISA), has seen increasing popularity in a wide range of applications. Like the radioimmunoassays, there are two basic types of ELISA: (1) direct binding assays primarily for detection or quantitation of antibodies; and (2) competition enzyme immunoassays for the detection and quantitation of antigens.

In this module ELISAs (the direct binding assay type) will be used to (1) observe a specific interaction between an antigen (BSA) and its antibody (anti-BSA), and (2) determine by titration the amount (i.e. titer) of anti-BSA antibody in a serum preparation from a rabbit immunized with the BSA antigen. The whole experiment will require two 3-hour lab periods and will include the following steps. In the first lab period: (1) prepare antigen solutions and coat the antigens to a 96-well microtiter plate; and (2) prepare serial dilutions of rabbit serum solutions. In the next lab period: (1) add antibody solutions to antigens coated on the plate to allow antigen-antibody interaction; (2) add second antibody conjugated with an enzyme (e.g. horseradish peroxidase enzyme conjugated with goat antibodies specific for rabbit antibodies); this second antibody, referred to as anti-Ig, will react with the antigen-antibody complex that formed earlier; and (3) add a substrate solution that will react with the enzyme conjugate to form a colour product. Theoretically, the intensity of colour developed (i.e. absorbance or optical density) is proportional to the quantity of anti-BSA antibody in the antiserum sample. No significant colour change will be observed if no anti-BSA antibodies are present in the rabbit serum.

Safety Guidelines

Some of the chemicals are irritants and could cause harmful reactions. Before proceeding be sure to put on lab coat, safety goggles, and plastic gloves. Avoid skin/eye contact with any reagent and do not ingest. Flush spills and splashes with water for at least 15 minutes. Report any accident to the instructor. At the end of the experiment each day, dispose of waste by following laboratory guidelines and wash hands thoroughly before leaving the laboratory.

Standard lab safety guidelines for handling chemicals and biological samples are adequate here. Students must wear lab coat (or apron), gloves and goggles in performing the experiment. The substrate solution which contains TMB, a potential mutagen, and organic solvent should be handled with caution.

Experimental Outline

Lab Period 1

(a) Pre-lab discussions of background information and protocol

(b) Coating antigens to microtiter plates

(c) Preparation of serial dilutions of rabbit sera

Lab Period 2

(a) Pre-lab discussion

(b) Saturation of antigen plate

(c) Addition of first antibody

(d) Addition of second antibody-enzyme conjugate

(e) Addition of substrate

(f) Read results

Briefly, the four basic steps of the assay are:

1. *Binding the Antigen*. BSA or gelatin is added to separate wells of the microtiter plate. Allow time for the antigen to bind to the plate.
2. *Antibody Addition*. Plates are washed. Rabbit serum A containing antibody to BSA and rabbit serum B (normal rabbit serum without antibodies to BSA) are serially diluted and added to specific wells of the ELISA plate. This is incubated to allow for the antigen-antibody interaction.
3. *Conjugate Addition*. The plates are again washed. This will wash away antibody that has not reacted with the antigen on the plate. The conjugate (enzyme coupled to an anti-Ig or second antibody) is added. Conjugate A binds rabbit antibody. Conjugate B binds to mouse antibody. Both conjugates are linked to horseradish peroxidase in this case. Again the plates are incubated.
4. *Addition of Substrate*. Again, wash away unbound material. An enzyme substrate (a colour forming chromogen in this case) is

added. This reacts with the conjugate to produce a colour indicating an antigen-antibody reaction.

Materials

PBS. Phosphate buffered saline (PBS) is provided as a 10X solution. Add 50 ml to a 500 ml graduated cylinder. Fill to the 500 ml mark with distilled H_2O for working solution. (Optional: add 0.5 ml of Tween 20 per liter of 1X PBS).

PBS (3% milk). Blocking diluent is made by adding 3 grams nonfat dry milk to 100 ml 1 X PBS.

Carbonate buffer 0.05M. Sodium carbonate buffer, pH 9.6. containing 0.02% sodium azide.

Antigen I 1% Bovine Serum Albumin (BSA). A vial is provided containing 0.1 grams of BSA. This must be mixed with 10 ml of carbonate buffer for working solution Antigen I.

Antigen II 1% Gelatin. Prepare a 1% solution by dissolving 0.1 grams into 10 ml of carbonate buffer. This must be heated until the gelatin is dissolved. Cool before using.

Rabbit Serum A. Contains antibodies against BSA. Provided in vial is 0.1 ml of a 1:10 dilution. Dilute this with 9.9 ml of PBS (3% milk) as follows. Empty contents of vial into a 15 ml test tube. Measure out 9.9 ml of PBS (3% milk). Use a small amount to rinse out vial. Empty this and remaining diluent into the 15 ml tube. Mix. Label tube #1 Rabbit Serum A 1:1000.

Rabbit Serum B Normal Rabbit Serum. Does not contain antibodies against BSA. Prepare as above for Rabbit serum A. Label #8 Rabbit Serum B 1:1000.

Anti-Rabbit Ig Conjugate. 0.2 ml provided. Keep frozen until ready to use. Dilute in 19.8 mls PBS (3% milk) in 50 ml tube. Label Rabbit Conjugate.

Anti-Mouse Ig Conjugate. Horseradish peroxidase conjugated antibody specific for mouse Ig. Store and dilute as for anti Rabbit Ig conjugate. Label Mouse Conjugate.

TMB Substrate. This is provided as solution A and B. Do not prepare until just before use. Mix 30 mls solution A with 30 mls of solution B. Label TMB. Keep from light.

Spectrophotometer, with microcuvette for 250 μl, ELISA Microplate Reader

96 well Microtiter plate

Pipetman, P 200

Multichannel pipet (optional)
Disposable pipet tips
Water, deionized or distilled
Bovine Serum Albumin (BSA) Solution, Antigen I
Gelatin Solution, Antigen II
Phosphate buffer saline (PBS), 10X
PBS 1 X with 3% milk, blocking solution and serum diluent
Carbonate buffer, antigen coating solution
Rabbit anti-BSA antiserum, rabbit serum A, 1:10 dilution
Anti-rabbit Ig conjugate, 1:50 dilution
Anti-mouse Ig conjugate, 1:50 dilution
TMB substrate solutions A and B (commercially available from Sigma Chemical Co)
96 well microliter plate (Immulon 1, Dynatech Laboratories, Inc. or equivalent)
Pipetman, P-200
Multichannel pipet (optional)
Disposal pipet tips
Plastic squirt bottle
Test tubes, glass, 12 x 75
Test tubes, plastic, 15 ml and 50 ml
Test tube racks
Vials or microcentrifuge tubes
5 ml pipets
Vortex mixer (optional)
ELISA microtiter plate reader (BIO-TEK Instruments or equivalent); optional and can be replaced by spectrophotometer with Microcuvettes

Pre-lab Preparation

1. Phosphate-buffered Saline (PBS), 10X
   2.3 g $NaHPO_4$ (anhydrous) (1.9 mM)
   11.5 g Na_2HPO_4 (anhydrous) (8.1 mM)
   90.0 g NaCI (154 mM)
   Add distilled H_2O to 800 ml
   Adjust to desired pH (7.2 to 7.4)
   Using 1 M NaOH or 1 M HCI
   Add HP to 1 liter

2. Carbonate buffer

   1.6 g Na_2CO_3 (15 mM)

   2.9 g $NaHCO_3$ (35 mM)

   0.2 g NaN_3 (3.1 mM)

   Add distilled H_2O to 1 liter

   Adjust to pH 9.5

   Caution: Sodium azide is poisonous; wear gloves.

3. PBS (1X) with 3% milk

   Dissolve 3 g nonfat milk powder in 100 ml PBS. Prepare during the day of use.

4. 1% Bovine serum albumin solution, (Antigen I)

   Dissolve 0.1 g BSA (e.g. from Sigma Chemical) in 10 ml carbonate buffer.

   Alternatively, provide a vial with 0.1 9 BSA.

5. 1% Gelatin solution, (Antigen II)

   Dissolve 0.1 g gelatin (e.g. from Sigma Chemical) in 10 ml carbonate buffer.

   Alternatively, provide a vial with 0.1 g gelatin.

6. Rabbit anti-BSA antiserum 1:10 (Rabbit, Serum A)

   Mix 0.1 ml of rabbit anti-BSA antiserum (e.g. from Sigma Chemical) with 0.9 ml of 1 X PBS. Prepare 0.1 ml aliquots in vials and freeze.

7. Normal rabbit serum 1:10 (Rabbit serum B).

   Prepare as above. Keep aliquots (0.1 ml) frozen.

8. Anti-Rabbit Ig Conjugate, 1:50

   Obtain goat anti-rabbit Ig conjugated with horseradish peroxidase (e.g. from Jackson ImunoResearch Lab, Inc.)

   Mix 0.1 ml with 4.9 ml 1 X PBS with 3% milk. Prepare 0.2 ml aliquots and freeze.

   (Adjust dilution according to batch and vendor's suggestions.)

9. Anti-Mouse Ig conjugate 1:50

   Obtain rabbit anti-mouse Ig conjugated with horseradish peroxidase and prepare dilution as above. Keep aliquots (0.2 ml) frozen.

10. TMB Substrate Solutions*

    Solution A

    In a one liter volumetric flask dissolve 1.0 g Tetramethyl benzidine (TMB) (Sigma Chemical) 600 mls methanol. Bring to 1.0 liter with glycerol. Store at 4°C in foil wrapped bottle.

Solution B

In a one liter volumetric flask dissolve in 500 ml distilled H_2O

22.82 g K_2HPO_4

19.2 g Citric Acid

1.34 mls 30% H_2O_2

Bring to 1.0 liter with distilled H_2O. Thimerosal (0.01 %) may be used as a preservative.

* Also available ready-to-use from Sigma Chemical

Method

Lab Period 1

Antigen Application to Plate

1. Prepare antigen I (BSA) and antigen II (gelatin) as described in C-6.
2. Copy the chart below. The numbers and letters on the chart correspond to the wells in the ELISA test plate. Along the sides of the chart record the materials added to each well as you continue with the procedure.
3. Using a pipetman, add 100 ml of Antigen I to wells in rows A, B, C, and D, columns 1 through 5 and 8 through 12 of the ELISA plate. Leave columns 6 and 7 empty to separate the two sections. (If a pipetman is not available, use a dropping pipet and add 3 drops of antigen solution.)
4. Using a clean pipet tip, add 100 ml of Antigen II to wells in rows E, F, G, and H, columns 1 through 5 and 8 through 12 of the ELISA plate. Record on your chart what was added to each row.
5. Seal the ELISA plate with plastic wrap and incubate at 37°C for 1 hour. (Room temperature Incubation is acceptable.) Refrigerate overnight.

Antibody Dilution in Tubes

6. Prepare stock solutions Rabbit Serum A (label tube #1) and Rabbit Serum B (label tube #2) as described in C-6.
7. Label 8 15 ml test tubes: 2, 3, 4, 5, 9, 10, 11, 12.
8. Using a 5 ml pipet, add 4 ml of phosphate buffered saline with dry milk (PBS:3% milk) to each of the 8 test tubes.
9. Using test tube 1 containing Rabbit Serum A, perform a serial dilution using a 5 ml pipet as follows:
   (a) Transfer 4 mls of rabbit serum A from tube 1 to tube 2. Mix well by drawing liquid up and down several times.

(b) Transfer 4 ml from tube 2 to tube 3. Mix well.
(c) Transfer 4 ml from tube 3 to tube 4. Mix well.
(d) Transfer 4 ml from tube 4 to tube 5. Mix well.

10. Using tube #8 containing rabbit serum B, repeat the serial dilution as described above for tubes 9 through 12. Seal all of the tubes and refrigerate until needed (step 14).

Lab Period 2

Antibody Addition to Plate

11. Empty contents of ELISA plates into a sink by turning upside down and flicking. Blot on paper towels.
12. Fill each well on the ELISA plate with PBS: milk to the top of the wells. Let set for 15 minutes. Empty as above. (This step serves the purpose of saturating the nonspecific binding sites in each well.)
13. Wash by flicking the plate and refilling wells with PBS only. Let sit for 2 minutes. Empty and blot as above. Repeat.
14. Using a pipetman, add 100 ml of diluted rabbit serum to each well according to the chart below. The tube number corresponds to the row on the ELISA plate. Use a clean pipet tip for each rabbit serum (A and B). Record all information. (If a pipetman is not available, use a dropping pipet and add 2 drops of each serum preparation.)
15. Cover plate and incubate for one hour at room temperature.
16. Empty ELISA plates into sink and blot on paper towels.

ELISA Plate Set Up

Tube #	*Rabbit Serum*	*Dilution*	*Column*	*Row*
1	A	1:1000	1	A - H
2	A	1:2000	2	A - H
3	A	1:4000	3	A - H
4	A	1:8000	4	A - H
5	A	1:16000	5	A - H
8	B	1:1000	8	A - H
9	B	1:2000	9	A - H
10	B	1:4000	10	A - H
11	B	1:8000	11	A - H
12	B	1:16000	12	A - H

17. Wash as in step 13. Wash a total of three times.
18. Add 100 ml of Rabbit Conjugate to wells A and B, E and F, rows 1 through 5 and 8 through 12. Add 100 ml of Mouse Conjugate to wells C and D, G and H, rows 1 through 5 and 8 through 12. Cover plates. Record information on chart.
19. Incubate for 15 minutes at 37°C

Substrate Addition to Plate

20. Prepare TMB substrate as described in C-6. Working solution of substrate should be *freshly* prepared right before use.
21. Empty plates and wash as described in steps 16 and 17 above.
22. Add 200 ml of fresh substrate to wells in rows A, S, C, D, E, F, G, and H, columns 1 through 5 and 8 through 12. A colour change (colourless to blue) after few minutes indicates a positive antigen-antibody reaction.
23. Read optical density or absorbance at wavelength 650 nm on a ELISA plate reader after 15 and 30 minutes. Alternatively, transfer the reaction product from each well to a microcuvette and read OD_{650} in a spectrophotometer. If neither a plate reader nor a spectrophotometer is available, place ELISA plate on an index card or sheet of white paper and observe any colour change. Make observations between 15 and 30 minutes and score according to the following:

    +++ very reactive

    ++ moderately reactive

    \+ slightly reactive

    \- no reaction
24. Record your results on your chart.

Results

1. Prepare a table as shown below and summarize your results. Give a title to your table.

	BSA		*GELATIN*	
	Rabbit Conjugate	*Mouse Conjugate*	*Rabbit Conjugate*	*Mouse Conjugate*
Anti-BSA				
1 : 1000				
1 : 2000				
1 : 4000				
1 : 8000				
1 : 16000				

Normal
Serum
1 : 1000
1 : 2000
1 : 4000
1 : 8000
1 : 16000

2. Using the data in your table above, prepare a graph to compare the activities of anti-BSA and normal serum control. Give a title to your figure. (Which format, table and figure, do you think is more effective in showing your results?)
3. Write a paragraph to interpret your data. It is important here to consider the results you obtain with both your test system (i.e., anti-BSA, BSA and Rabbit Conjugate) and controls (normal serum, gelatin and mouse conjugate). What can you say about the specificity and potency (i.e., titer) of the anti-BSA preparation?
4. Discuss any problem you may have encountered in your experiment as well as any discrepancies you may have found in your results.
5. Try to draw a conclusion of your experiment. Do you think you achieved the goal of your experiment? Explain.

Results Analysis

1. A typical and expected result is shown in the chart below. Actual result may vary according to the reagents used.

	|←Rabbit Serum A→| Antigen I (1% BSA)					|← Rabbit Serum B→|						
	1	2	3	4	5	6	7	8	9	10	11	12
R-A	+++	++	++	+	-			-	-	-	-	-
R-B	+++	++	++	+	-			-	-	-	-	-
M-C	-	-	-	-	-			-	-	-	-	-
M-D	-	-	-	-	-			-	-	-	-	-
Antigen II (1% gelatin)												
R-E	-	-	-	-	-			-	-	-	-	-
R-F	-	-	-	-	-			-	-	-	-	-
M-G	-	-	-	-	-			-	-	-	-	-
M-H	-	-	-	-	-			-	-	-	-	-

+++ = very reactive M = mouse conjugate
++ = moderately reactive
+ = slightly reactive R = rabbit conjugate
– = not reactive

1:4000 = the highest dilution of anti-BSA antibody that gives a visible antigen-antibody reaction.

2. The end result of an ELISA test may be subjectively assessed by eye to provide a "yes'' or "no" answer. Positive results are indicated by a colour reaction which can be distinguished from the colourless (or very faint colour) wells of negative controls.

   Preferably, the coloured reaction product should be read in a ELISA microplate reader or a spectrophotometer to measure more precisely its absorbance (i.e. optical density) at the maximal absorption wavelength. A threshold value is determined (usually the upper limit of absorbance value shown by the negative controls) and test samples giving an absorbance value above the threshold level (e.g. 2 to 3 x negative control) is considered as positive.

   Since the speed of enzyme reaction may vary the end result should be read as soon as possible, and every 5-10 minutes thereafter for 30 minutes. If the reaction appears too fast it can be terminated by adding 50 ml of hydrofluoric acid (HF, 1:400 dilution) to each well.

   The result may be expressed as either endpoint titration or a quantitative O.D. measurement, or as both.
3. The procedure will work for most commercial reagents (e.g. the antigens, rabbit sera and enzyme conjugates) but the concentrations may need to be adjusted by following vendor's recommendations.
4. Incubation times with antibody can be reduced to 30 minutes if speed is important although some loss of binding may occur. Incubation at 37°C will speed the reaction but may also increase nonspecific binding leading to high background
5. No positive results - sources of problem
   (a) BSA antigen: check concentration and date of antigen solution; check plates that are used for coating; check antigen coating buffer.

   (b) Rabbit Anti-BSA serum - check date and dilutions of antibody solutions; increase concentration of antibody; call supplier for advice.

   (c) Anti-rabbit Ig conjugate: check date and dilution of conjugate; increase concentration of conjugate; call supplier for advice.

   (d) TMB substrate: Prepare fresh working solution; check stock solutions A and B.

   (e) Check (c) and (d) by mixing 100 ml of diluted conjugate with TMB working solution. Colour reaction should result.
6. High Background - sources of problem

(a) Plate not-saturated: check saturation buffer (PBS milk); prolong saturation time; increase milk concentration; try other carrier proteins such as 10% normal goat serum.
(b) Reaction too fast or end results read too late: try reading the results soon after substrate addition; stop reaction with HF.
(c) Conjugates too concentrated: Dilute conjugates further; call supplier for advice.
(d) Sera too concentrated. Dilute antiserum and normal serum further; call supplier for advice.

8

ANTIGEN-ANTIBODY REACTIONS

The combination of antibodies with corresponding antigens is the basis of most immunologic reactions, both for serum antibodies and for lymphocyte membrane receptors. Study of the antigen-antibody reaction is complicated by the fact that most antigens are macromolecules and that, even when their structure is known, the nature, number, and localization of the various functional groups present on each molecule are difficult to assess. For this reason, haptens with a limited number of determinants are used to study antigen-antibody reactions. Extrapolation to more complex molecules is then attempted. This approach is warranted, because although antigens and haptens differ in their immunogenicity that is their capacity to stimulate antibody production, they share the ability to combine with antibodies, which is our present interest. In practice, one may generalize the discussion with the use of the term "ligand" to indicate either antigen or hapten.

Reactions of Antibodies with Haptens

Antibodies specific for simple haptens may be obtained with a fair degree of purity quite easily. Haptens are coupled to carrier proteins for the purpose of immunization or are utilized directly, as for oligosaccharides or polysaccharides that contain the hapten in their sequence. The antibodies thus obtained are purified by passage of the antiserum through an immunoadsorbent that contains the hapten, followed by elution of specific antibodies.

Antibody Valence and Affinity

In the classic example, a hapten (or ligand) with only one functional group is bound reversibly to the specific antibody site. For a given antibody concentration, as increasing amounts of hapten are mixed with antibody, the concentration of bound hapten progressively increase until all antibody sites are occupied. When saturation is achieved, the number of bound hapten molecules equals the number of antibody sites. The number of sites on each antibody molecule defines the antibody's valence. If one supposes that binding sites are identical for all antibody molecules of the antiserum and that they react independently of each other, the reaction may be represented as follows:

$$A + H \underset{k}{\overset{k}{\rightleftharpoons}} AH$$

where A is the antibody site, H is the haptenic site, and k and k' are the rate constants for association and dissociation of the antibody-hapten (AH) complex.

If one makes the hypothesis of identity and functional independence of the sites, at equilibrium, the mass action law is written:

$$[A][H]^{K} = [AH]^{k'}$$

or

$$\frac{[AH]}{[A][H]} = \frac{k}{k'} = K.$$

The ratio K = k/k' is the intrinsic association constant of the reaction. It quantitates the stability of the hapten-antibody complex or, more precisely, the intrinsic affinity of the antibody site for the hapten. The parentheses corresponds to respective concentrations of antibody sites occupied by the hapten [AH], of free antibody sites [A], and of free haptenic sites [H]. One notes that the units of the K ratio are concentration divided by concentration squared. Therefore, its dimensions are the reciprocal of concentration, most often in units of liters per mole. Increased intrinsic affinity of the complex is associated a larger K value,

Affinity Measurement

If in the preceding equation K = [AH]/[A][H], one knows the total concentration of antibodies present in the reaction, [A] + [AH], which is the sum of bound and free antibody concentrations, and also the concentration of hapten, [H] + [AH], the sum of bound and free hapten; to calculate K, it is sufficient to determine [A], [H], or [AH] at equilibrium. Of these three parameters, the easiest to measure is usually [H], the free hapten concentration. Several methods allow separation of free and bound hapten. The most used method is equilibrium dialysis, which is satisfactory, because it does not alter the equilibrium previously reached by the antigen-antibody reaction, as may theoretically happen when immunoglobulin is precipitated by concentrated saline solutions. Other methods include Ig precipitation by ammonium sulfate or polyethylene glycol, inhibition of precipitation by hapten addition, ultracentrifugation, and fluorescence quenching. Equilibrium dialysis will be dealt with in more detail, because the principles of this method are also applicable in other settings.

Equilibrium Dialysis

The technique is based on the fact that free hapten, H, of low molecular weight is dialyzable, whereas antibody-hapten complex, AH, is not. Antibodies and hapten are placed into two compartments separated by a membrane. The hapten of low molecular weight diffuses freely across the dialysis membrane, whereas the antibody molecules, whether free or complexed with the hapten, do not cross the membrane. At equilibrium, the hapten concentration is higher in the compartment that contains the antibodies, because the free hapten is in equal concentration on both sides of the membrane. Figure shows the evolution of hapten concentration on both sides of the membrane. In the absence of antibodies, both concentrations are the same at the time of equilibrium. In the presence of antibodies, the difference between concentrations in the two chambers corresponds to hapten that is bound to antibodies. In theory, a difference in concentrations between the two chambers might exist independently of binding to a specific antibody because of Donnan's effect of negatively charged proteins. This phenomenon may be excluded by adding a sufficient concentration of salt to the medium, that is, 0.15 M sodium chloride. Obviously, Donnan's effect does not pertain to electronegative hap tens. It is noteworthy that the same equilibrium is obtained when the antibody and the hapten are initially placed in one chamber or in separate chambers, which provides good evidence for the reversibility of antibody-hapten binding.

Simple measurement of the total hapten concentration on both sides of the dialysis membrane permits measurement of the association constant, K. If c represents the free hapten concentration (measured experimentally in such units as moles per liter in the antibody-free chamber), r is the average number of hapten molecules bound per antibody molecule at equilibrium, M is the antibody concentration (mole/liter), and if n equals the maximum number of hapten molecules that each antibody molecule can bind (i.e., antibody valence),

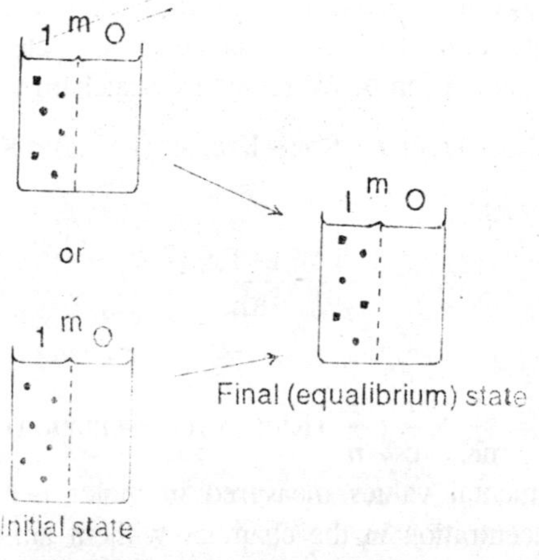

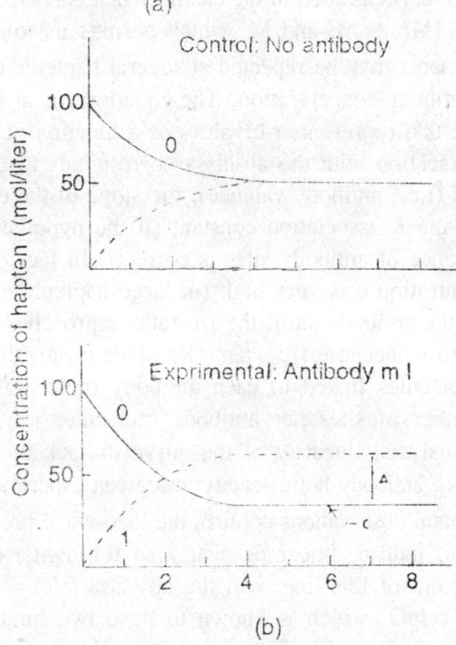

Fig. 8.1. Equilibrium dislaysis. (a) Low-molecular-weight haptens (small dots) diffuse freely between compartments I and O, but antibody molecules (large dots) cannot diffuse. At equilibrium, the hapten concentration is higher in I, due to binding by antibody. (b) Changes in haptenic concentrations with time. Equilibrium is reached in about 4 hr.

$$K = \frac{[AH]}{[A][H]} = \frac{Mr}{M(n-r)\times c} = \frac{r}{(n-r)c}$$

where c is an experimental value (mole/liter), n is a constant, and r equals an average number of sites experimentally determined by Mr. The preceding equation can be expressed by Scatchard's representation:

$$\frac{r}{c} = Kn - kr, \text{ or } r = Knc - Krc, \text{ or } r(1 + Kc) = Knc$$

or, by rearrangement,

$$\frac{1}{r^0} = \frac{1 + Kc}{Knc}$$

alternatively

$$\frac{1}{r} = \frac{1}{nK} \times \frac{1}{c} + \frac{1}{n} \text{ (Klotz's representation).}$$

The experimental values measured in moles per liter, are the free hapten concentration in the chamber without antibodies [H], c; the total hapten concentration in the chamber that contains the antibodies [AH] + [H], (Mr + c), and M, which permits us to calculate r.

Measurements may be repeated at several haptenic concentrations for a fixed antibody concentration. The equation r/c = Kn - Kr states that the curve that represents r/c values as a function of r is a straight line, the intersection with the abscissa corresponds to the number of antibody sites (i.e., antibody valence), the slope of the equation is the reciprocal of the *K* association constant (if the hypothesis of identity and independence of antibody sites is correct). In fact, when the free hapten concentration c is very high (in large haptenic excess close to saturation of the antibody site), the r/c ratio approaches zero, and r is nearly equal to n, because r/c - Kn - Kr. This means that the number of hapten molecules linked to each antibody molecule approximates the total number of sites per antibody molecule, i.e., valence. By Scatchard's analysis, linearity of the curve or lack of it also allows an estimation of antibody homogeneity in a given experimental situation.

Experimental observations confirm the theoretical prediction. Curve r/c = f(r), is, indeed, linear or nearly so for high r/c values. The intersection point of this line with the abscissa (r/c = 0) gives an r value of 2 for IgG, which is known to have two binding sites, and generally r = 5 for IgM. One may also note from the family of curves, that high-affinity antibodies (hyperimmunization) show a straight line with a steeper slope than do antibodies of low affinity obtained after short immunization with high antigen doses. Experimentally, K values range between 10^6 and 10^{12} liter/mole, depending on antibody affinity.

Affinity Heterogeneity: Average Affinity

The r/c = f(r) curve is not linear in the majority of experimental situations, which indicates that most antisera contain antibodies with sites that are not independent or not identical. The bivalence of IgC molecules is not in question, because the same types of curves are seen with monovalent IgG fragments prepared by enzymatic digestion. This excludes potential interactions between free and occupied sites on the IgG molecule. In fact, the nonlinear nature of the curves is explained by the heterogeneity of antibody affinity. The only homogeneous antibodies are those directed against certain polysaccharides and myelomatous antibodies. The affinity of these antibodies is sufficiently homogeneous so that the curves r/c = f(r) are straight lines. With these exceptions, the term "antibody" used in the singular represents most often a heterogeneous molecule population that has in common the capacity to bind one given hapten or antigen. In the usual case of affinity heterogeneity, an average intrinsic association constant K_0 is defined, which is the average of the association constants for various antibodies.

Antibody heterogeneity may be quantitated if one hypothesizes that association constants of the various antibodies are normally distributed (with a gaussian distribution) around the K_0 average. This is done by Sips' equation: $r/(n - r) = (Kc)^a$, where a is a dispersion index of association constants similar to the standard deviation of normal distributions (when antibodies are homogeneous, a = 1). Sips' equation may be written:

$$\log\frac{r}{n-r} = a \log K + a \log c.$$

log r/(n – r) is thus a linear function of log c.

For r = n/2, log Kc = 0, Kc = 1, finally, K_0 = 1/c, which shows that the average intrinsic association constant is independent of *a*.

The average intrinsic association constant is highly variable for different antigen antibody systems. It is usually of the order of 10^5 to 10^7. One must understand that when multiple sites are present, the actual affinity may be much higher than the intrinsic affinity, because the valence augments the stability of the complex by cooperation between sites. This cooperation applies particularly to antigen molecules that have numerous determinants. This phenomenon of monogamous binding may increase 10^3- to 10^5-fold the effective association constant. Thus, the association constant of antidinitrophenyl (anti-DNP) antibodies is 100,000 times higher for DNP bound repetitively to bacteriophages than for the monovalent hapten derivative DNP-lysine (10^2 to 10^7). Monogamous binding decreases greatly the risk of dissociation. This fact probably explains the remarkable biologic activity in vivo of very low amounts of antisera directed against viruses and bacteria, which often bear repetitive determinants.

Affinities of the various antibodies present in a given serum are not always distributed in a gaussian way. In some cases, the antibodies may show marked homogeneity. In a case described by Nisonoff, a rabbit anti-*p*-azobenzoate antibody was obtained in crystallized form. The presence of a limited number of molecular species may sometimes explain the observed heterogeneity. Analysis of Sips' curves obtained with an anti-DNP rabbit antibody, and with a wide spectrum of free hapten concentrations by Werblin and Siskind, has shown very limited antibody heterogeneity in some cases or, at least, no gaussian distribution.

The affinity of IgG antibodies for monovalent haptens is generally higher than is that of IgM antibodies for the same specificity. Similarly, IgG affinity for antigenic erythrocyte sites is more than 100 times greater than the affinity of Fab fragments (3.5 S) produced by enzymatic cleavage of native IgG with papain. This difference in affinity could be due to binding of 7S molecules on red blood cells by their two antibody sites. The intrinsic affinity of an isolated antibody-combining site is usually much lower than the functional affinity measured with the entire molecule. Thus, in the system cited above, one may compare intrinsic and functional affinities of IgG and IgM anti-DNP antibodies to DNP-lysine, and to a preparation of DNP coupled to phage $\phi \times 174$. The functional affinity of anti-DNP IgG measured by phage neutralization is 10^3 times higher than the intrinsic affinity measured with the monovalent ligand. In the IgM case, the ratio is of the order of 10^6. This difference probably reflects the higher valence of IgM molecules and their capacity to form multiple bonds with structures of sufficient density of antigenic determinants as discussed earlier.

Specificity and Affinity

Immunologic reactions derive their uniqueness from their specificity; specificity essentially defines antigens and antibodies. We have seen that even a purified antiserum contains usually a mixture of antibodies of variable affinity. The specificity of an antibody is linked to the ability of this antibody to discriminate between antigenic determinants of similar but different structures, by binding with different affinities; the greater the affinity difference of an antibody for two given hap tens, the higher the specificity of the antibody for the hapten of higher affinity. This is reminiscent of the early Landsteiner definition of specificity: "a disproportional action of several related antibodies toward related haptens." The high affinity antibodies sometimes seem less specific than do low-affinity antibodies when the antibodies are present in low concentrations, because they combine less well with the homologous hapten, but, in fact, the affinity difference between haptens is essential, more so even than are absolute affinity values and even more so than the binding intensity. In practice, antibody *specificity* is generally defined in relation to an antigen or hapten family (the antibody affinity for several antigens or haptens is measured), Antibody *affinity* is, in general, defined with respect to an antibody family (more compares the average affinity of several antisera for a given hapten).

Physicochemical Basis of Hapten-antibody Binding

The forces that link antibodies to haptens or to antigens are not fundamentally different from those govern the various interactions between proteins or between enzymes and substrates. The nature of these forces and the thermodynamics of the reactions in question are thus similar to those of interactions between proteins. They are exclusively non covalent forces and therefore relatively weak and highly dependent on steric complementarity between antigen and antibody sites.

Forces involved in Antigen-antibody Reactions

The forces that link antigens and antibodies are of four types: Van Der Waals forces, electrostatic forces (or coulombian forces), hydrogen bonds, and hydrophobic bonds. Their intensity is variable: Van Der Waals forces are the weakest, of the order of 0.5 kcal/ mole *versus* 2 to 5 kcal/ mole for electrostatic forces or hydrogen bonds. They are, in any case, of lesser strength than covalent forces, which reach 50 to 150 kcal/mole. It is important to note that these forces are at least decreased proportionally to the reciprocal of the increasing distance between antigen and antibody molecules. They are stronger at shorter distances, that is, when the two structures are complementary. Increased stability of the antigen-antibody combination is seen when the contact area widens. The theoretical distance between antigen and antibody sites may also be influenced by the flexibility of the antibody molecule, which may allow certain residues to approach closer to the contact area with the antigen.

Van Der Waals Forces

These forces are linked to the electron movement within molecules and lead to the rapid formation of electric fields that polarize surrounding molecules. The presence of electric fields within one molecule (dipole formation) changes the distribution of electrons in the other molecule, and the two dipoles then attract each other. The interactional energies are inversely proportional to the sixth power of interatomic distances. These relatively weak forces are functional only for closely complementary antigens and antibodies and in these cases are very efficient.

Electrostatic Forces

Electrostatic (or coulombian) forces are those exerted between two ionic groups of opposite charge, for example, the amino (NH_3^+) group of a protein lysine and the carboxyl group (COO^-) of aspartate from another protein.

Hydrogen Bonds

These bonds are of weal energy (0.5 to 2 kcal/mole) between electropositive (hydrogen) and electronegative atoms (oxygen or nitrogen). Interactional energies are inversely proportional to the sixth power of the distance between the atoms. The stability of these bonds depends on the presence of water molecules; they are augmented by decreased temperature.

Hydrophobic Bonds (Stabilizing Interactions between Nonpolar Groups in Water)

Certain amino acids of Ig chains tend to repel solvent water molecules and do not establish hydrogen bonds in aqueous solution. This phenomenon is seen with aromatic amino acids, such as valine and leucine. However, when two of these amino acids come in close contact, after a diminution in the number of neighbouring water molecules, there is an energetic increment that augments the bond stability. The energetic increment is higher as the distance between two amino adds decreases. In contrast to hydrogen bonds, the hydrophobic bonds between antigen and antibody are more stable at higher temperature.

Conclusions

Once the antigen (or hapten) and the antibody have made contact, several links form between molecules; the different forces discussed above may be involved to varying degrees. The nature of the bonds indicates that binding between antigen and antibody is reversible under

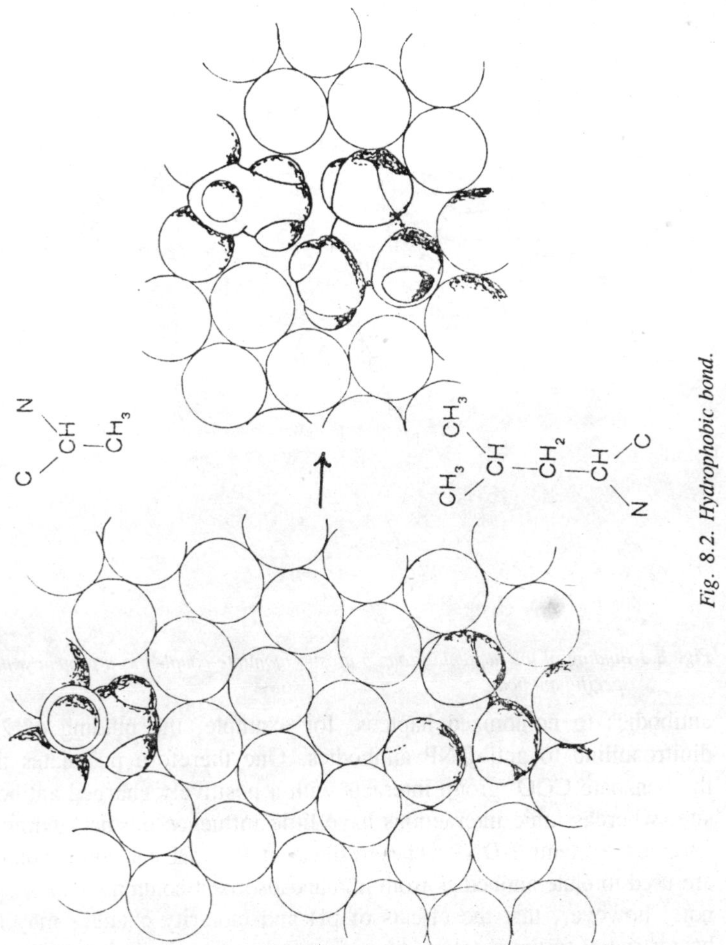

Fig. 8.2. Hydrophobic bond.

thermal agitation of molecules due to heating; the bond is more stable when structure are more complementary.

Influence of Physicochemical Conditions on Antigen-antibody Reactions

Effects of pH and Ionic Strength

Antigen-antibody reactions are affected by the existing physicochemical conditions. The effect of these conditions on association constants depends on the nature of the forces involved. Thus, the binding of *p*-aminobenzoate and antibenzoate antibodies decreases from 0.1 to 1 M NaCl. Conversely, pH change does not affect the binding of

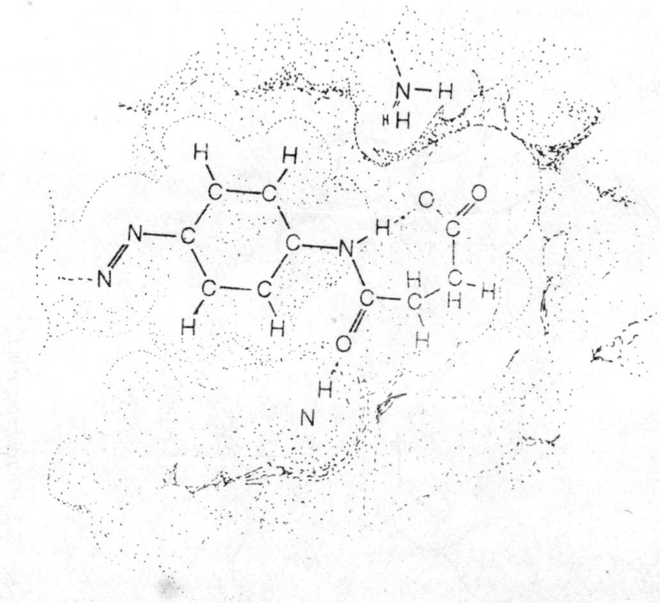

Fig. 8.3. Binding of the haptenic group p-azosuccinonilate coupled to a protein with its specific antibody.

antibodies to nonionized haptens, for example, the binding of 2.4-dinitroaniline to anti DNP antibodies. One therefore postulates that the benzoate COO^- group interacts with a positively charged antibody site, whereas ionic interactions have little influence on the binding of antibody to neutral DNP. These effects of low pH and high molarity are used to elute antibodies from immunoadsorbent columns. One should note, however, that the effects of pH and molarity changes may not be restricted to the reactive sites of antigen or antibody but may alter all of the molecule.

Effects of Temperature

The binding forces between antigens and antibodies vary with temperature, which affects the rate of the antigen-antibody reaction. This rate may be optimal at 37°C, especially for some "secondary" reactions, for example, precipitation or complement fixation. The association constant is not necessarily higher, but the reaction occurs more rapidly at 37° than at 4°C. Some reactions occur more rapidly at 4°C. In other cases, the speed of association is no different at 1° and 40°C.

Thermodynamics

The formation of an antigen-antibody complex causes a loss of free energy, ΔF°, proportional to the logarithm of the average association constant:

$$F^0 = - R \times T \times \ln K_0$$

where *R* is the constant of perfect gas (1.987 cal/mole), *T* equals the absolute temperature. In K_0 represents the Neper logarithm of the association constant, and $\Delta F°$ is the variation of standard energy (p = 1 atm; T = 293), that is the energy gain or loss (calories) observed when 1 mole of antibody reacts with 1 mole of hapten. $\Delta F°$ values vary from 6000 to 11,000 cal for association constants of the order of 10^5 to 10^9 M^{-1}, at 20° or 30°C.

Kinetics

Kinetics of antibody-hapten reactions are generally studied with low hapten concentrations under conditions such that reactions are slow enough to be measured. It is also possible, using a special apparatus, to measure variations in free hapten concentrations at microsecond intervals, thus permitting utilization of higher hapten concentrations. Finally, relaxation methods may be used, in which antibody and hapten are brought to equilibrium, and then equilibrium is modified by change in temperature, pressure, or electric field, and the return to original equilibrium is examined.

The association reaction between a small hapten and an antibody is very rapid, no more than 10 times less than simple diffusion (10^8 liter/mole/sec). Reactions between whole proteins and antibodies are also rapid, but they are 10 to 100 times slower than those that occur between simple haptens and antibodies. It is important to recognize that association kinetics of haptens and antibodies are only slightly dependent on the intrinsic association constants. Therefore, affinity differences are linked primarily to variations in dissociation rates.

Reactions of Antibodies with Macromolecules

All complexes of antibodies and monovalent haptens are soluble. Macromolecular antigens, however, may show precipitation in the presence of certain antibodies. The precipitation reaction has broad biologic significance, because practically all antibodies are precipitating under certain conditions. For this reason, in addition to its historic interest, we will discuss precipitation reactions in detail, making a distinction between theoretical considerations (which essentially concern liquid precipitation) and practical applications (which usually utilize gel precipitation). Before discussing precipitation, we will review the major theoretical aspects of reactions between antibodies and macromolecular antigens, extrapolating insofar as is possible from the data on haptens reported above.

Valence and Complexity of Protein and Polysaccharidic Antigens

Macromolecules, as indicated, present multiple antigenic determinants. The number of determinants per molecule is often unknown. These determinants may be identical and repetitive (for example, for polysaccharides) or may be different (as for proteins). In both instances, the practical result is that antigen is complex and multivalent. Extrapolation of the data obtained from study of haptens to macromolecular antigens is therefore of limited usefulness.

Affinity and Avidity

Because most antigens are multivalent, it is generally impossible to define the intrinsic association constant K for a given antibody. In fact, association forces of antibody and antigen molecules do not depend only on association constants but also relate to the number of bonds per antigen molecule. We have seen earlier that the monogamous binding phenomenon, in which several sites of a given antibody molecule react with two closely related determinants of the same antigen molecule, may increase by 10^5 times the intrinsic association constant of an antibody for a homologous monovalent ligand. For a given complex antigen, the number of functional groups is usually unknown, and even purified antisera contain a mixture of antibodies. Finally, nonspecific factors, such as ionic strength, may also interfere. This fact explains why one often, in practice, utilizes the term "avidity", which simply refers in a semi-quantitative way to the ability of an antiserum to exert certain biologic functions that relate directly to binding to specific antigens, such as agglutination or viral neutralization, reserving the term "affinity" for intrinsic association constants.

Measuring affinity for antibodies directed against macromolecular antigens is obviously more difficult than for antihapten antibodies. One may employ Scatchard's analysis and Sips' equation, as has been previously described for antihapten antibodies. However, one must be aware of the limits of these methods, in which numerous nonverified hypotheses are involved. An approximation of affinity may be obtained by examining concentration of bound antigen at different antigen to antibody ratios (the curve being steeper for high affinity antibodies), by diluting the mixture of antiserum and antigen (which reduces free antigen and antibody concentrations and leads to dissociation) or by adding cold antigen to the complexes formed by antibodies and labeled antigen. The sera with the lowest affinity show the earliest dissociation. The precipitation curve also gives information, because the curve is steeper for high-affinity antibodies. Differences in avidity between antisera may be due to (1) differences in the average intrinsic affinity constant of antibodies, (2) varying "specificity", that is, antisera that contain differing proportions of antibodies directed against the various functional groups on the antigen, (3) different proportions of antibody molecules that give rise to the monogamous binding phenomenon, and (4) different numbers of combination sites, (lgM contain five or 10 sites per molecule and therefore are of higher affinity than IgG, with two sites).

Classification of Antigen-antibody Reactions

Numerous methods have been described to detect and to measure interactions between antibodies and antigens. These methods belong to three categories:

Primary Methods

These methods involve the primary interactions between antigen and antibody, independently of biochemical and biologic phenomena that may follow. Primary reactions depend only on the quantity and affinity of antibodies. The proportion of antigen-bound for a given antibody concentration is very sensitive to changes in antigen concentration when the antibody is of high affinity but much less sensitive when antibodies are of low affinity. Conversely, the absolute amount of bound antigen is more variable with respect to antigen concentration for low-affinity than for high-affinity antibodies. This difference indicates that primary-type reactions simultaneously evaluate antibody quantity and affinity, thus the interest in each case of studying several antigen concentrations.

Secondary Methods

These methods relate to phenomena that are the direct (but not obligatory consequence of primary interactions. Precipitation is one example: some antibodies give rise to precipitates, but not all antibodies are precipitating, at least under certain technical conditions. Similarly, IgM molecules may be very easily detected by agglutination reactions, whereas the agglutination reaction is 50 to 500 times less sensitive for IgG. Complement fixation and hemolysis, which require CIq binding, ignore noncomplement fixing antibodies like IgA and IgF. Secondary methods do not necessarily indicate the entire amount of antibody present in an antiserum. The methods are, nonetheless, very useful in studying heterogeneous antibody mixtures. They may even be used for quantitation in well-defined systems.

Table 8.1. Antigen-antibody Reactions

Primary reactions
Farr test (ammonium sulfate, polyethylene, glycol, anti-Ig sera)
Equilibrium dialysis
Fluorescence quenching
Immunofluorescence
Radioirnmunologic dosage
Secondary reactions (in vitro)
Precipitation
Agglutination
Neutralization (toxins, virus, phages)
Complement-dependent reactions (cytolysis, phagocytosis, chemotaxis)
Tertiary reactions (in vivo)
Intravascular hemolysis
Anaphylaxis (anaphylactic shock, passive cutaneous anaphylaxis)
Arthus' reaction

Tertiary Methods

These methods examine the in vivo (biologic) consequences of primary interactions. Their complexity is even greater than that of secondary methods. In addition to the variables inherent in the primary and secondary reactions, other parameters specific to the individual, such as complement level or cellular receptors for IgG Fc fragments, must be considered. The limitations concerning secondary reactions thus also apply to tertiary reactions.

Direct Binding Tests

Primary binding tests have common features. They require the utilization of purified antigen or antibody. Methods used to measure antigen and antibody concentrations must be quantitative and sensitive. They usually utilize isotopic or fluorescent labeling. They all involve the separation of soluble antigen-antibody complexes from free antigen or antibody molecules.

Methods that Employ Precipitation of Antigen-antibody Complexes

Under certain conditions immune complexes may precipitate, whereas antigen alone does not. Thus, 50% ammonium sulfate

Table 8.2. Quantitative Methods that Allow Dosage of Antibodies, Based on the Direct Binding between Antigen and Antibody

Methods
Methods that depend on the separation of bound and free antigen by precipitation of the antigen-antibody complex
Ammonium sulfate
Polyethylene glycol (mol wt 6000)
Heterologous anti-Ig serum
Methods that depend on fluorometric measurement
Fluorescence quenching
Fluorescence augmentation
Fluorescence polarization
Methods that depend on differences in size of bound and free antigen molecules
Equilibrium dialysis
Ultracentrifugation
Gel filtration
Methods that depend on differential electrophoretic mobility of bound and free antigen
Paper or acetate electrophoresis
Polyacrylamide gel electrophoresis
Radioimmunologic assays

precipitates Ig (whether free or complexed) without precipitating *bovine serum albumin* (BSA). The use of labeled BSA allows one to easily measure the proportion of bound BSA at different dilutions of anti-BSA antiserum. This statement describes the Farr test, which is performed at various ratios of labeled antigen and antiserum. It is of practical importance to know that the addition of ammonium sulfate does not itself alter the proportion of labeled BSA bound to antibody which means that association or dissociation of immune complexes is not affected. Results of the Farr test are expressed as the percentage of bound antigen at a given antiserum dilution or, better yet, by the antiserum dilution that gives 33% precipitation, which is the dilution that corresponds to antigen excess and soluble immune complexes This method has very wide application. It is commonly used for albumin the M protein of streptococci, somatic antigens of gram-negative bacteria, and DNA. If antigen alone is precipitated by ammonium sulfate, polyethylene glycol (PEG) may be substituted, as may anti-Ig sera (coprecipitation), which precipitate Ig but not antigen. For example polyvinylpyrrolidone is precipitated by 50% ammonium sulfate but not by 20% PEG. The sensitivity of the coprecipitation method is similar to that using ammonium sulfate, except for "early" IgM antibodies of low avidity. In the latter case, coprecipitation is less sensitive (due to cross-reactions of the test antibody with antigens present in the anti-Ig antiserum, nonprecipitated IgM or incomplete prevention of complex dissociation by anti-Ig antibodies). In addition, coprecipitation is obviously more expensive than ammonium sulfate.

Fluorometric Methods

Fluorescence quenching

Most proteins emit fluorescent energy in the ultraviolet spectrum. This energy is decreased when certain molecules are bound to the proteins. Fluorescence is due to phenylalanine, tyrosine, alanine and especially tryptophan, which emit at 345 nm. Ig molecules fluoresces. Their excitation is maximal at 290 nm, and the emitted fluorescence has a wavelength of 345 nm. Ig interaction with a hapten diminishes Ig fluorescence, because the hapten absorbs part of the energy, which is then either absorbed or emitted at another wavelength. In either case, emitted fluorescence at 345 mm is diminished. In practice, one measures the extinction of fluorescence after addition of several hapten concentrations.

The main advantage of the technique is its rapidity, simplicity, and the minimal amounts of material consumed. In addition,

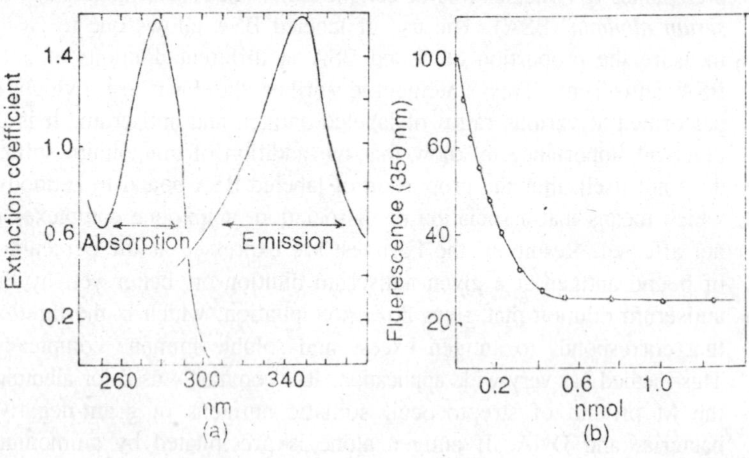

Fig. 8.4. Fluorescence quenching. (a) Absorption and emission spectra of purified anti-DNP antibodies. (b) Dimination of antibody fluorescence after binding to ε-DNP lysine.

nondialyzable haptens may be examined. Its limitations relate to the necessity of using purified antigen and antibody preparations. This purity is required because Ig fluorescence represents only a small portion of total serum fluorescence. In addition, it is necessary that the absorption spectrum of hapten overlap that of Ig. Furthermore, numerous haptens and sugars do not modify Ig fluorescence and can therefore not be studied by this method.

Fluorescence augmentation

Some molecules that do not fluoresce is aqueous solution do so after binding to certain proteins. Thus, when a fluorescent antigen is incubated in vitro with an antibody, its fluorescence (measured at 345 nm) may significantly increase (the physical basis of the increase is not completely understood). This method is little used, even though it is simple, sensitive, quantitative, and without a requirement for pure antibody preparations. It should be noted however, that only antigens that shows an increase in fluorescence may studied with this method.

Fluorescence polarization

The degree of polarization of the light emitted by fluorescent molecules (excited by polarized light) depends on the size of the molecule. Small molecules with little rotational movement give minimal polarization. If, however, the fluorescent trolecule is bound to a large particle, like an antibody, its rotational movement is restricted, and light polarization increases. Thus, the proportion of antigen molecules bound to antibody may be evaluated by measuring fluorescence polarization. This method can be applied only to small antigens and haptens, the fluorescence spectrum of which (natural or modified by the introduction of fluorescent substitutes) is different from that of Ig. It is a simple, rapid, and sensitive technique; it gives information about the association constant and antibody heterogeneity and eventually on the kinetics of the antigen-antibody reaction. Here also, purified antibodies must be used.

Methods Based on Size Differences between Bound and Free Antigen

Equilibrium dialysis

This method has been presented in detail in connection with haptens. It is applicable to antigens of a molecular weight less than 10,000 daltons.

Ultracentrifugation

Antibody-antigen complexes sediment more rapidly by ultracentrifugation than do free antigen molecules. By use of radiolabeled antigen, one may obtain bound and free antigen at different sedimentation levels that correspond to various antigen-antibody ratios. The dilution of antiserum associated with a given ratio of free to bound antigen can be evaluated. Unfortunately, this technique is not too precise and is not easily applied to mixtures of IgG and IgM antibodies or to antigens larger than 7S IgG molecules.

Gel filtration

Antigen-antibody complexes travel more rapidly in Sephadex than do free antigen molecules and may thus be separated. This method has been used in measuring anti-insulin antibodies.

Other Methods

Electrophoretic methods (on paper, acetate, or polyacrylamide gel) are based on the differential electrophoretic migration of free antigen (for example, BSA or diphtheria toxoid) and the immune complex. By use of labeled antigens, one may then calculate the percentage of free or bound antigen (radio counts are determined in several sections of paper or gel).

Another technique consists of measuring the reactivity of a purified-labeled antiserum in the presence of a nonlabelled antigen. This method has been used to quantitate erythrocyte rhesus antigens after elution of red blood cell antiserum. At the present time, the precision of this technique has not been established.

Radioimmunoassays

Radioimmunoassays were initially developed as sensitive and specific methods for measuring levels of circulating hormones. The method depends on the binding of a radiolabeled hormone to a standard antiserum of high specificity and avidity for the hormone in question. Admixture of the two reagents produces soluble immune complexes that may be dissociated by unlabelled hormone. The displacement of labeled hormone at different concentrations of unlabeled hormone is determined and a standard curve is established by plotting the ratio of bound to free hormone as a function of added unlabelled hormone concentration. An analogous curve is then prepared by use of the reference standard (cold hormone) and the test serum. By comparing the curves the serum level of the hormone in question can be calculated.

The methodologic problems posed by radioimmunoassays are similar to those discussed previously for the Farr test, particularly in separating bound and free antigen. Separation may be facilitated by taking advantage of the phenomenon of spontaneous binding of some antigens to inert surfaces (paper or glass). Antigen-antibody complexes, in contrast with

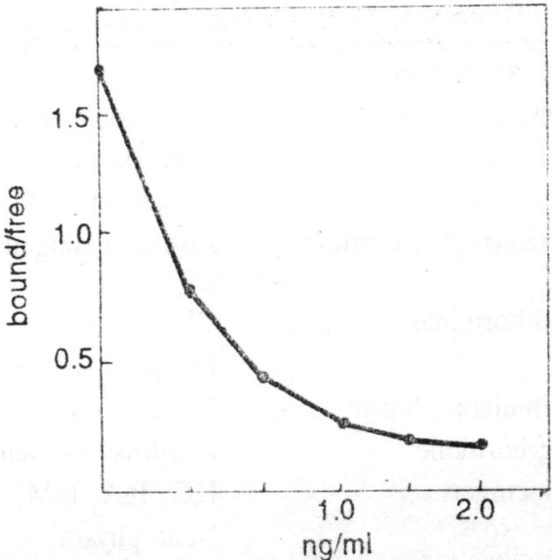

Fig. 8.5. Radioimmunoassay. Standard curve for the evaluation of a peptide hormone.

free antigen, do not bind to these surfaces and are therefore recoverable. Examples of this separation include the adsorbance of insulin onto paper; certain peptide hormones (insulin, growth hormone, parathormone, and ACTH) adsorb onto glass, talc, silica, or Fuller's earth. Similarly, some compounds, such as insulin, bind to carbon. For example, silica earth is added to tubes that contain a mixture of antibodies, antigens, and immune complexes. The mixture is agitated, the tubes centrifuged, and the supernatant decanted. The radioactivity that remains in the tube is that of the free hormone bound on silica earth. Conversely, techniques have been developed in which the antibody (not the antigen) is bound onto inert particles. It is possible to bind antigrowth hormone antibody to polystyrene or polypropylene tubes or to Sephadex-derived beads without destroying antibody activity. The antigen may then bind to the antibodies fixed onto the tube walls or onto Sephadex beads.

Generally speaking, radioimmunoassays are rapid, easily reproducible, very precise, and sensitive methods. Their only limitations relate to the difficulty of obtaining certain reagents, such as labeled hormones, and antisera directed against antigens or haptens of small size, and also to the risk of cross-reaction between serum antigen and the test hormone. This risk is diminished, but not eliminated, by using very pure antibodies and hormones.

Immunofluorescence

Immunofluorescence, discovered by Coons in 1941, consists of detecting antigenic determinants present in cells or tissues by application

Table 8.3. Main Radioimmunoassays

Assays that use an antibody as the reagent	
Peptide hormones	Nonhormonal substances
Insulin	Intrinsic factor
Growth hormone	cAMP, cGMP, cIMP, cUMP
Adrenocorticotrophic hormone (ACTH)	Australia antigen (and antibody)
Parathyroid hormone	CI esterase
Glucagon	Fructose 1,6-diphosphatase
Thyroid-stimulating hormone	Carcinoernbryonic antigen
Luteinizing hormone	Rheumatoid factor
Placental lactogen	IgG, IgA, IgM, IgF.
Proinsulin	Neurophysin
Secretin	Thyroid-binding globulin
Vasopressin	α-Fetoprotein
Angiotensin	Plasmin
Oxytocin	Ferritin
Bradykinin	Prothrombin
Thyroglobulin	Tetanus antibody
α-MSH	Botulic toxin
β-MSH	Staphylococcal enterotoxin
Gastrin	A and B fibrinopeptides
Calcitonin	Fibrin
Proinsulin C-peptide	Fibrin degradation products
PZ-CCK	
T3, T4	Drugs
Prostaglandins	Amphetamines
Erythropoietin	Barbiturates
	Digoxin, digitoxin
Nonpeptide hormones	D-Tubocurarine
Testosterone	Hydantoin
Estradiol	Lysergic acid diethylamide
Adosterone	Medroxyprogesterone
Estrone	Methylprednisolone
Dihydrotestosterone	Morphine
	Penicillin
	Phenothiazines
	Vitamin A

Nonimmune assays systems that use a substrate different from antibody but that also use saturation analysis

Hormones	Specific reagent
Thyroxin Cortisol Corticosterone Cortisone 11-Deoxycortisol Progesterone Testosterone	Specific plasma-binding proteins
Nonhormonal substances	**Specific reagent**
Vitamin B_{12}	Intrinsic factor
Folic acid	FA reductase
Cyclic AMP and GMP	Kinase protein
Messenger RNA	Complementary DNA

of a specific antiserum previously made immunofluorescent by the binding of fluorochrome. After washing the preparation, only the antibodies specific for an antigen present in the preparation remain bound and may be detected after illumination by ultraviolet light.

Fluorochromes are substances that emit a higher-wavelength (530 nm) light than the excitor light (490 nm). The two products most commonly used are fluorescein isothiocyanate (green fluorescence) and rhodamine isothiocyanate (orange fluorescence). Ig labeling must preserve antibody activity. Purified antibodies are combined with fluorochrome: the fluorochrome excess is eliminated by dialysis or filtration can Sephadex gel. The reading requires a microscope illuminated by ultraviolet light emitted by a mercury vapor lamp. The initial filter (excitator) eliminates the visible rats and allows ultraviolet to pass. A second filter (stopping filter) placed between the objective and the ocular eliminates ultraviolet light and allows passage only of fluorescent light.

In practice, one may label directly the antigen-specific antibody. Direct immunofluorescence is most commonly used in studying tissue sections. One may also initially bind the nonlabelled antibody onto the structures studied and then detect binding by addition of heterologous fluorescent anti-Ig serum (indirect immunofluorescence). A third technique consists of adding the specific antibody and complement and then detecting complement binding (thus, antibody binding) by addition

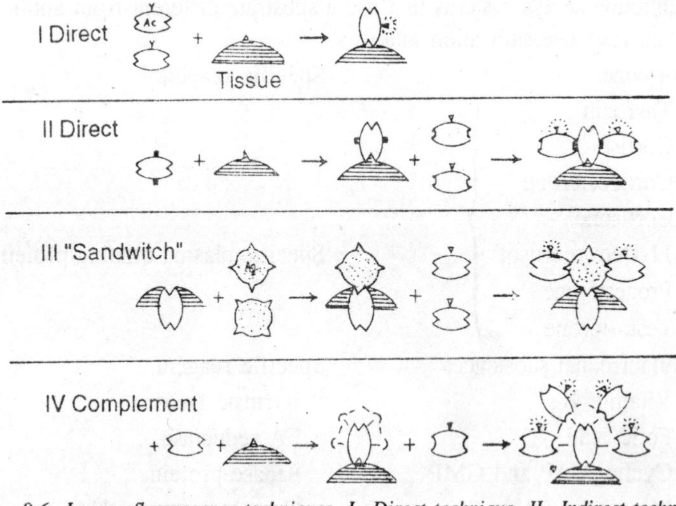

Fig. 8.6. Immunofluorescence techniques. I—Direct technique. II—Indirect technique. III—Sandwich technique. IV—Complement technique.

of a fluorescent anticomplement serum. Immunofluorescence methods are widely used in virology, bacteriology, parasitology, and immunopathology for detecting circulating antibodies to human tissue antigen (e.g., antibodies to nuclear mitochondria, glomerular basement membrane, gastric cells or smooth muscle).

Immunoprecipitation

When a specific antiserum and the antigen are placed into a small-diameter tube, a precipitate forms at the interface of the two solutions. The ring test may be used for qualitative immunologic studies. It represents the simplest example of the immunoprecipitation phenomenon, the theoretical and practical applications of which have played a considerable role in immunology.

Liquid Precipitation

Immunoprecipitation was discovered in 1897, but widespread application awaited the introduction of quantitative methods by Heidelberger in 1933. Heidelberger initially showed that a major antigen of the pneumococcal capsule was a polysaccharide. Conceptually this discovery was important because it established the immunogenicity of nonprotein substances (polysaccharides). Even more importantly, the nonprotein nature of the antigen, for the first time, permitted separate measurement of antigens and antibodies within an immune complex.

Description of the Quantitative precipitation reaction

If the capsular antigen or type-III pneumococci (SSS III) and the antiserum produced by a rabbit immunized against this antigen are mixed together, a precipitate is observed at certain antigen to antibody ratios. The reaction is specific, because no precipitate forms when the antiserum is replaced by rabbit serum collected before immunization, or if another antigen is substituted for SSS-III antigen. After repeated washing, analysis of the precipitate reveals the presence of only proteins and polysaccharides. After elution of complexes, proteins appear to be Ig entirely precipitable by type-III polysaccharide. The eluted polysaccharides can once again be precipitated by an antipneumococcal antiserum. Thus, all protein present, measured by Kjeldahl technique, in the precipitate are antibodies (with the minor exception of complement traces).

When antigen concentration is increased, keeping antibody concentration constant, the amount of precipitated protein increase, reaching a maximum value and thereafter paradoxically decreasing.

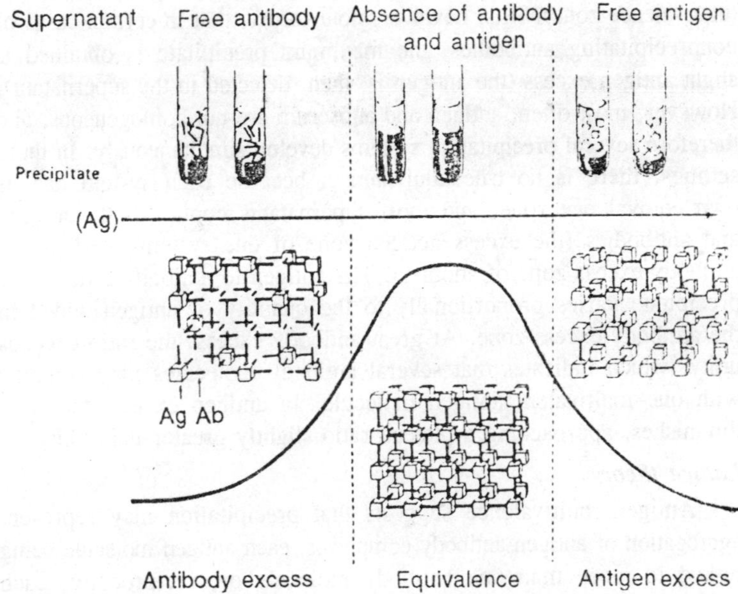

Fig. 8.7. Quantitative precipitation. Increasing amounts of antigen are added to a constant amount of antibody. The supernatant may contain free antibodies in excess (left) or free antigen in excess (right). The supernatant does not contain antigen or antibody at equivalence (middle).

The existence of an optimal quantity of precipitate for a very precise antigen to antibody ratio may be simply explained by the hypothesis that the number of antibody sites occupied by the antigen is regularly augmented until a maximum, when a lattice is formed. Once saturation is reached, the additional antigen produces partial lattice dissociation. We shall examine later the theoretical basis for the prozone phenomenon. These early observations concerning saccharidic antigens have general application to protein antigens.

Prozone Phenomenon

The content of free antigen or antibody in supernatant may be very easily analyzed after centrifugation and elimination of precipitate. At this point, one may simply add new antigen to detect free antibody or, alternatively, add fresh antibody to detect free antigen. If antigen or antibody are homogeneous, that is, do not comprise a mixture with different specificities, it is noted that none of the tubes contain both free antigen and antibody. One finds free antibody in the ascendant zone (antibody excess), free antigen in the descendant zone (antigen excess), but neither free antigen nor antibody at maximum precipitation (equivalence zone). It is, however, noteworthy, that in certain cases of nonprecipitating antibodies, the maximum precipitate is obtained is slight antigen excess (the antigen is, then, detected in the supernatant). However, most often, antigen and antiserum are not homogeneous, and therefore several precipitation systems develop simultaneously. In these settings, there is no true equivalence, because each system has its own equivalence zone, and most supernatants contain both antigens and antibodies (the excess antigen zone of one system overlaps the antibody-excess zone of another). The antigen to antibody ratio in the precipitate varies proportionally to the quantity of antigen added in the antibody-excess zone. At great antibody excess, the ratio exceeds unity, which indicates that several antibody molecules may combine with one multivalent antigen molecule. In antigen excess, the ratio diminishes, approaching a plateau ratio slightly greater than unity.

Lattice theory

Antigen multivalence suggests that precipitation may represent aggregation of antigen-antibody complexes, each antigen molecule being linked to more than one antibody molecule or, reciprocally, each antibody molecule being linked to more than one antigen molecule. When the size of aggregates exceeds a specified threshold, they spontaneously become insoluble. The hypothesis implies that antibodies are multivalent, as has been demonstrated by equilibrium dialysis.

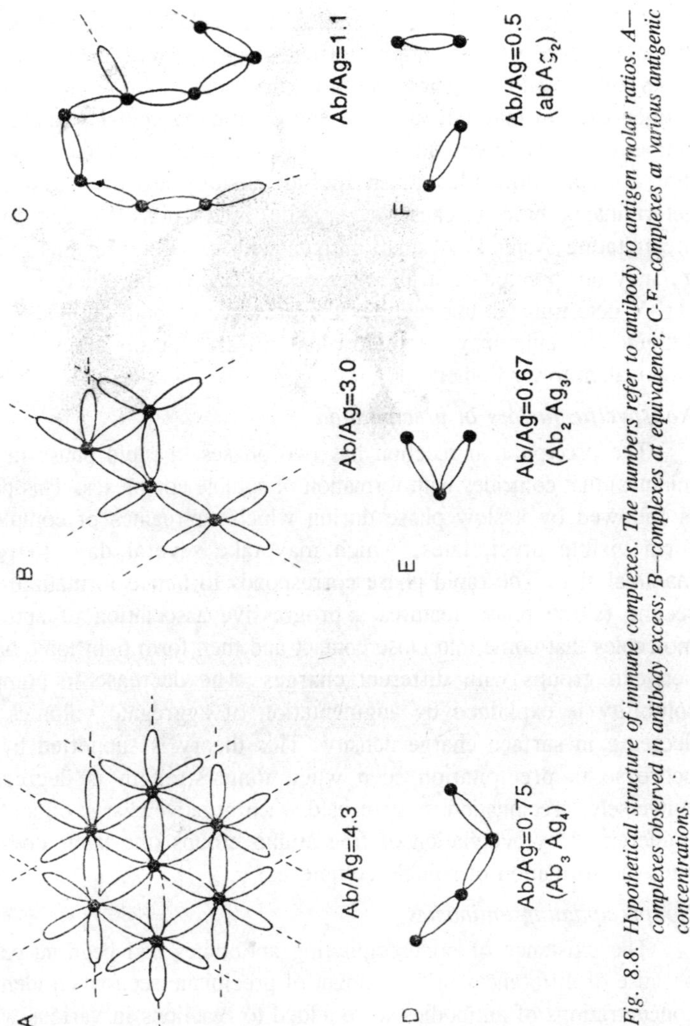

Fig. 8.8. Hypothetical structure of immune complexes. The numbers refer to antibody to antigen molar ratios. A—complexes observed in antibody excess; B—complexes at equivalence; C-F—complexes at various antigenic concentrations.

The lattice theory explains variations in the antibody to antigen ratio observed in the precipitates that were long at variance with the dogma of fixed composition of chemical compounds. In large antigen excess, the antibody to antigen ratio approaches unity, defining complexes of linear structure and regular antigen-antibody sequences. In antigen excess, complexes have antibody to antigen ratios of less than 1: 0.75 [Ab_3Ag_4] in weak antigen excess, a ratio of 0.67 [Ab_2Ag_3] with great antigen excess, and 0.5 [$AbAg_2$] in very high antigen excess. These ratios apply to bivalent antibody molecules. The formation of the latter small complexes, which do not precipitate, explains the disappearance in antigen excess of precipitate formed at equivalence. Consequently, when antibody to antigen ratios are very high, antibody valence (in antigen excess) may be estimated and/or minimal antigen valence (in antibody excess) approximated. Analysis is complicated by the fact that antigen molecules often include several different antigenic determinants that may cause (synergistic) interactions between several precipitating systems. A small antigen with two determinants, X and Y, may not precipitate in the presence of one or the other anti-X or anti-Y determinants but requires the presence of both antibodies. An antigen molecule may be multivalent toward certain antibodies but monovalent toward others.

Nonspecific factors in precipitation

The precipitation reaction has two phases: a rapid phase (a few minutes) that coincides with formation of soluble complexes. This phase is followed by a slow phase during which aggregates of complexes form visible precipitates, which may take several days to reach maximal size. The rapid phase corresponds to lattice formation. The second (slow) phase features a progressive association of antibody molecules that come into close contact and then form tight ionic bonds between groups with different charges. The decrease in complex solubility is explained by augmentation of aggregate volumes and decrease in surface charge density. This theory is supported by the decrease in precipitation seen when ionic strength is decreased; conversely, precipitation is augmented when negative charge of antibody is increased by acetylation of free amino groups (the latter does not alter the formation of soluble complexes).

Nonprecipitating antibodies

The existence of nonprecipitating antibodies had been suspected because of differences in the amount of precipitate seen when identical concentrations of antibodies were added to reactions in various ways. As an example, if one adds to an antiserum one-teeth of the amount of antigen that gives equivalence, a precipitate forms. Additional precipitate forms when new antigen is repeatedly added. The sum of all precipitated antibodies is much lower than the amount of antibodies directly precipitated in the equivalence zone: the difference represents non precipitating antibodies, that is molecules that do not in themselves precipitate but are included in complexes that contain precipitating antibodies of the same specificity. It was previously thought that these nonprecipitating antibodies were monovalent. They are, however,

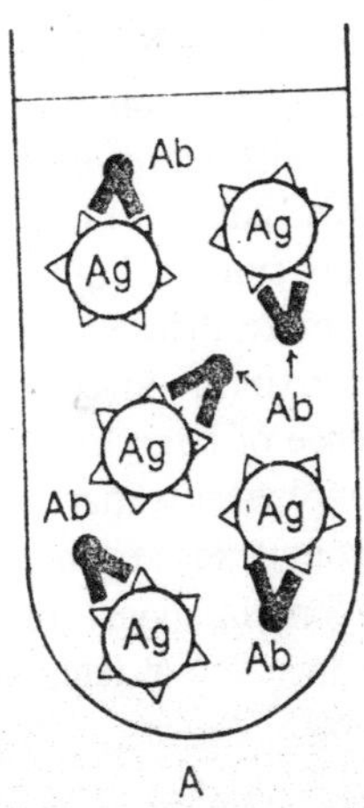

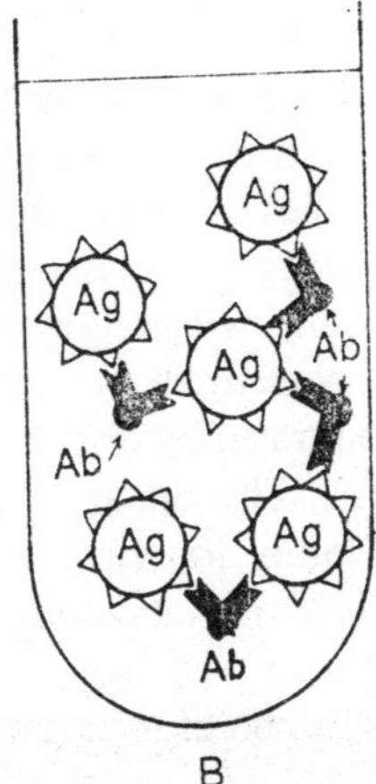

Fig. 8.9. Monogamous binding. An IgG antibody molecule may bind by its two sites to the determinants of the same antigen molecule (A), which very much increases solidity of the binding but prevents formation of the lattice normally observed (B).

multivalent but of low affinity. High antigen concentrations must be added to obtain precipitation, thus the formation-in antigens excess of soluble complexes.

Let us also note that certain high-affinity antibodies may not precipitate, because they preferentially combine with two determinants of the same antigen molecules to form cyclic complexes. This phenomenon may take place when the determinants are closely repetitive on an antigen surface (for example, in certain hapten-protein conjugates and for polysaccharides located on the surface of bacteria, viruses, or red blood cells). Some antibodies may, then, block precipitations of antigens by other antibodies. Because of their relative inflexibility of antibody molecules, monogamous binding develops and prevents the binding of one antibody molecule to two antigen molecules. Nonetheless, monogamous binding considerably increases the association constant.

Precipitation reversibility

Precipitates already formed are dissociated by adding fresh antigen (antigen excess). The reversibility is linked to transformation of large complexes into smaller ones. One apparent exception to this phenomenon, the Danysz phenomenon, deserves mention: when diphtheria toxin and antitoxin antibodies are mixed in equivalence, neutralization of the toxic activity depends on how mixing is performed. If toxin and antiserum are added simultaneously, neutralization is complete. But, if the toxin is added on two or three occasions at approximately 30-min intervals, neutralization is incomplete. This observation is explained by formation of indissociable complexes that contain more antibodies than antigen molecules, which leaves insufficient antibodies to neutralize the toxin secondarily added. In fact, all complexes are dissociable, even though some contain high-affinity antibodies with a very slow dissociation rate (more than 30 min, and even sometimes several days). Complex reversibility may also be demonstrated by displacing antigen in the presence of large free antigen excess. The apparent irreversibility of complexes is particularly noteworthy for viral and antiviral antibody complexes, even at high dilution (which lower free antigen and antibody concentrations), probably because of the increase in association constants that relates to the phenomenon of monogamous binding (vide supra).

Flocculation

The precipitation curve, which express the amount of precipitate as a function of the antigen concentration, does not always have the aspects. In some cases, particularly with horse sera, the formation of insoluble aggregates (precipitation) is only observed after addition of relatively large amounts of antigen. Here, antibody excess inhibits precipitation, as opposed to the more usual inhibition by antigen excess. Precipitation is observed in a very narrow range of antigen to antibody ratios. This phenomenon is called flocculation. Flocculation is observed with certain horse sera, particularly antisera directed against diphtheria

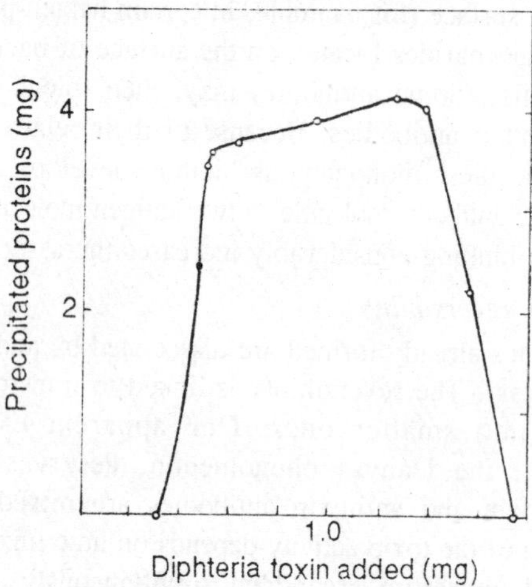

Fig. 8.10. Flocculation reaction. Increasing amounts of diphtheria toxin are added to constant amounts of antitoxin sera.

toxoid or streptococcal toxins. This property is not species specific, because horse sera directed against proteins or polysaccharides may give precipitation curves of the usual form. In addition, certain human sera may give flocculation with thyroglobulin. Flocculation mechanisms are linked to particular properties of antibodies rather than of antigens, because an antigen that flocculates with one antiserum will not show the same response with other antisera. Flocculation mechanisms are not completely understood. They may be provoked by formation of particularly soluble complexes due to high-affinity nonprecipitating antibodies (by the mechanism of, for example, monogamous binding). Only after these antibodies are saturated can the remaining antibodies precipitate the antigen. Flocculation is sometimes used to quantitate antitoxin the intense precipitation observed when one varies the ratio of antigen and antibody concentrations allows a macroscopic titration that is more precise than classic immunoprecipitation. Ramon's technique uses varying antibody concentrations and Dean and Webb's technique employs varying concentrations of antigen in which antigen concentrations are varied.

Gel Precipitation

Immunoprecipitation is of great historical importance because of the considerable expansion in knowledge it has produced, as suggested in the preceding discussion. A very wide spectrum of techniques of immunologic analysis has derived from it. The discovery of gel precipitation represented the critical stimulus. We shall first examine general rules of gel immunodiffusion, before reviewing major applications, including plate immunodiffusion, radial immunodiffusion, counter electrophoresis, and immunoelectrophoresis, all ingenious refinements on the initial technique of gel immunoprecipitation in tubes described by Oudin. When one antigen and one antibody are introduced into geified medium at different loci, they each diffuse, and a precipitate forms at the meeting point, if the ratio of antigen to antibody concentration is adequate. Depending on the support structure, one uses tube and plate immunodiffusions.

Tube Immunodiffusion (Oudin)

The first gel immunoprecipitation technique was described by au din at Pasteur Institute in 1946. The technique consists of placing agar gel into a glass tube that contains an antiserum, allowing solidification of the gel at 20°C, and then adding an antigen solution. The tube is kept vertical. The antigen spreads by simple diffusion throughout the gel, creating a concentration gradient (which decreases from top to bottom). When an adequate antigen concentration is achieved, a precipitate forms at the progression front of the antigen. In fact, the precipitate extends from the progression front to the zone in which the antigen concentration is excessive. In large antigen excess, the precipitate redissolves into soluble complexes, just as in liquid medium. It is observed that the precipitate line progresses toward the tube bottom, the progression front creating a stable precipitate as rapidly as the antigen concentration becomes sufficient. The upper portions of the precipitate resolve when the antigen concentration becomes too high. The distance traversed by the precipitate line depends on time (the distance covered is proportional to the square root of time), on the coefficient of antigen diffusion in gel, which essentially relates to molecular weight and shape of the molecule, and on antigen concentration in the upper reservoir.

The two last factors mentioned explain that when several antigens diffuse simultaneously in a gelified medium that contains antibodies directed toward each of them, precipitates are not superimposed, each

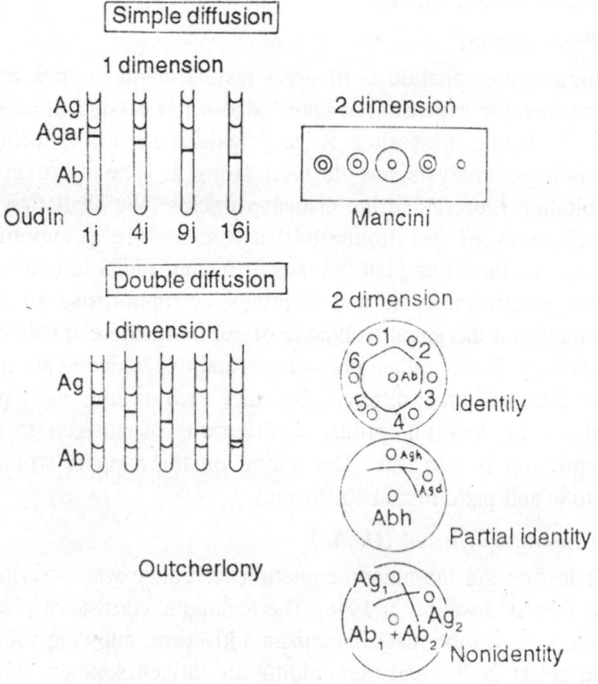

Fig. 8.11. Immunodiffusion techniques.

one migrating as its own speed. Obviously, it is possible that two distinct antigens have a similar migration speed and give rise to precipitates that cannot be differentiated. Finally, the number of lines permits estimation of the minimal number of antigens (or antibodies) present in the test solution.

It is possible to apply the tube immunodiffusion technique to determine antigen concentration under standardized conditions, because migration speed depends only on antigen concentration. One may, then, measure antigen and antibody concentrations in the precipitate by means of optical density (photometry). This immunodiffusion technique (tube) may be modified in several ways. Antigen may be incorporated into the gel and antibody added to the reservoir. This modification allows the study of flocculation, which, unlike precipitation, is inhibited by antibody excess. Another widely utilized modification is double immunodiffusion in tubes. In this technique, the gel that contains the antiserum is covered by gel without antiserum, and on top of these two gels either free antigen solution or agar-incorporated antigen is added. In this system, antigen and antibody diffuse toward each other in the neutral (gel) zone. Precipitates are formed at the point where antigen and antibody concentrations correspond to equivalence. The precipitate is then immobilized, and its intensity is augmented by accumulation of antigen and antibody molecules. Antigens and antibodies continue to react without a change in concentration ratio, at least for two reservoirs filled with amounts of antigen and antibody close to equivalence. In other situations, the precipitate slowly migrates away from the reservoir in which antigen or antibody is in excess.

Plate Immunodiffusion (Outcherlony)

The use of glass plates to substitute for immunoprecipitation in tubes represented major technical progress. Tubes are replaced by plates that contain wells. Antigen and antibody solutions placed into separate wells diffuse freely in gel and give rise to precipitates at equivalence. If the antibody concentration is in relative excess compared to antigen and if gel diffusion coefficients are equal, bands are formed close to the antigen-containing well; the opposite is observed in situations of antigen excess. The plate technique is simpler than the tube technique. It has, in particular, the advantage of allowing direct comparison of different antigen preparations. Various antigen preparations are placed into separate wells arranged in a circle, the center of which contains an antiserum-filled well. When the same antigen preparation is simultaneously placed into two adjacent wells, the two precipitation lines join and fuse if the wells are close together; this phenomenon is a "reaction of identity". If two different antigenic solutions are placed into two adjacent wells, the precipitation lines intersect, producing a "nonidentity reaction". Finally, if the two antigenic solutions give rise to cross-reactions and if one of them has been used to produce antiserum, the bands will fuse, but beyond the point of fusion, a projection is noted that extends the precipitate line formed by the immunizing antigen, yielding a "partial identity, or cross-reaction". This projection, which is generally less dense than the main precipitate, represents antibodies that react with antigenic determinants not shared by the two antigens. This reaction generally corresponds to a smaller quantity of antibodies than does the main reaction, which explains why the projection is not very dense and tends to be seen closer to the central well, where antibody concentrations are greater. The incarnation of this projection is clearer when antigen diffusion is faster (for example, because of low molecular weight). The precipitate is linear when the antigen and the antibody have approximately similar molecular weights; if the antigen is of higher molecular weight than is the antibody, the precipitate has a curve that is concave in the direction of the antigen (the converse is seen when the antigen has a lower molecular weight than does antibody). When there are determinants that give rise to cross-reactions between two antigens that were not used to prepare antisera, the projection is directed toward the antigen reacting less well with the antiserum, or there may be two distinct projections. These projections will have weak intensity and will be seen only if particularly careful inspection is

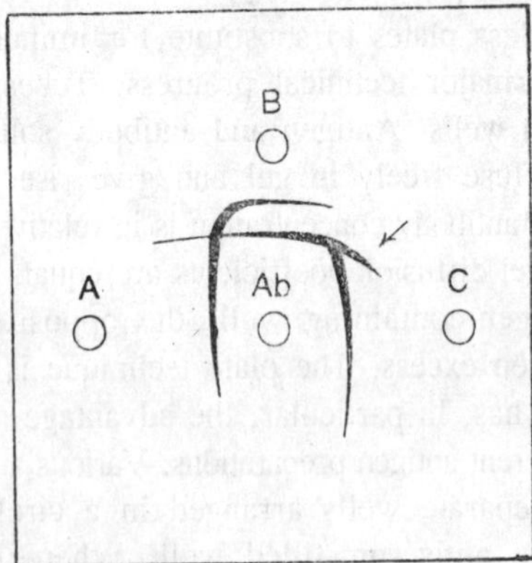

Fig. 8.12. Analysis of cross-reactions by Outcherlony's technique.

carried out. As in tube immunodiffusion, plate immunodiffusion permits analysis of antigen-antibody mixtures; the number of precipitates represents the minimal number of precipitating systems.

Radial Immunodiffusion (Mancini)

Agar into which a monospecific antiserum was previously incorporated is placed on a glass slide. The antigen solution deposited in a well diffuses into the agar, forming a blurred precipitate, the exterior diameter of which is proportional to the initial antigen concentration. This technique was developed by Mancini to quantitate certain antigens. Monospecific antiserum is used, and the ring diameters of several known amounts of antigen are recorded. Mancini's technique is routine, used to measure Ig and complement.

The use of low antiserum concentrations makes the assay more sensitive but does not allow evaluation of high antigen concentrations that correspond to antigen excess and therefore do not precipitate. For weak antigens (or low antigen concentrations), particularly IgD, it is possible to augment the precipitate by addition of a protein precipitating agent such as tannic acid. One may thereby detect amounts as minimal as 0.3 mg/100 ml of IgD. It is also possible to utilize radiolabeled antigens and quantity by autoradiography.

Immunoelectrophoresis (Grabar and Williams)

Another simple and discriminative technique of immunodiffusion is immunoelectrophoresis. The technique consists of introducing an antigen mixture into a well cut in an agar plate and applying an electric field for 1 to 2 hr to separate antigen molecules according to their electrophoretic mobility. The electric field is then discontinued, and a polyspecific antiserum is added to a groove that parallels the direction of migration of the antigen preparation. Antibodies and antigens are allowed to diffuse freely toward one another, giving rise to precipitates analogous to those described in Outcherlony's technique, with similar standards for interpretation. Immunoelectrophoresis has been particularly rewarding in detecting the presence of more than 30 proteins in normal human serum.

Radioimmunoelectrophoresis

The immunoelectrophoresis technique may be modified to study antibodies directed toward haptens (which do not give rise to visible precipitates) or to study small amounts of antibody. The antigens are placed into a well and fractionated by gel electrophoresis in the usual way. A groove is then prepared, and radioactive monospecific antiserum plus an anti-Ig antiserum are added. The antigen combines with the specific antibody, and the complexes are precipitated by the anti-Ig serum. The nonprecipitated proteins are eluted after washing and fixation, and the plates are read by autoradiography.

Crossed Electrophoresis (Laurell's First Technique)

Laurell has proposed another variant of immunoelectrophoresis that allows quantitative analysis of antigens. The technique begins like classic electrophoresis. The various antigens present in the gel are separated linearly. A narrow strip of gel that contains the fractionated antigen is cut out. This gel band is placed onto a new gel plate into which weak antiserum concentrations were previously incorporated. The electric current is then reapplied perpendicular to the direction of the first electrophoresis. Precipitation zones develop at each antigen-antibody front. Precipitation lines are higher for more concentrated antigens. Analysis of the various precipitates is easier than in classic immunoelectrophoresis. In addition, for a given antibody concentration, the position of the peak is grossly proportional to antigen concentration.

Rocket Immunoelectrophoresis (laurell's Second Technique)

A simple modification of the preceding technique permits easy and rapid measurement of antigen concentrations. A monospecific antiserum is incorporated into a gel slide. Different dilutions of the antigen preparation being tested are placed into parallel wells. An electric current is then applied perpendicular to the line of wells. A rocket-like blurring ("trails") forms and extends as far as the antigen is in excess. The existence of a linear relationship between antigen concentrations in the holes and precipitate heights leads to precise and sensitive measurement of antigen concentration.

Counterimmunoelectrophoresis

This technique is very similar to the previous method, in which the antiserum is place into a row of wells that parallel wells that contain antigens. A current is then applied perpendicular to the lines of wells, and the maximal antigen dilution (or antibody dilution) that gives a precipitate is noted, by comparison to controls of known antigen (or antibody) concentrations. This technique is particularly useful for the detection of antibodies that precipitate the antigens that do not migrate well, if at all, in agarose gel unless and electric field is applied.

Antigen-antibody Reactions using Cellular or Particulate Antigens

We have discussed in the preceding text various techniques for the study of reactions between antibodies and soluble antigens, including equilibrium dialysis, the Farr test, fluorescence quenching, and immunoprecipitation. It is also possible to study the reaction between antibodies and cellular or particulate antigens. These techniques, commonly used in serology, include agglutination reactions, neutralization reactions, and reactions that involve complement fixation (complement fixation, cytolysis and opsonization).

Agglutination Reactions

The agglutination reaction represents binding of agglutinating antibody to cells. It is particularly applicable to bacteria and erythrocytes. Red blood cells can be directly studied or used to support soluble antigens bound to their surface (passive hemagglutination).

Agglutination Mechanisms

Agglutinating antibodies agglutinate cells in a 0.15 M sodium chloride solution, as contrasted with nonagglutinating antibodies. This simplistic concept of agglutinating and nonagglutinating antibodies is only relative, however. The phenomenon depends on the molecular structure of the antibody (IgM or IgG), hence its valence, on the density of antigenic receptors on the cell surface, and on the medium employed (ionic strength). Generally speaking, agglutinating antibodies are IgM and nonagglutinating antibodies are IgG, but some IgG agglutinating antibodies exist. Nonagglutinating antibodies may be detected by artificial agglutinating techniques (Coombs test, enzymatic red blood cell treatment, centrifugation, and addition of macromolecular solutions).

Cell agglutination (for example, of erythrocytes or bacteria) is not necessarily immunologic. Nonimmunologic agglutination may occur in the absence of antibodies by addition of polycations, macromolecular substances, protamine sulfate, or glucose. Phytohemagglutinin provokes agglutination of erythrocytes by binding to receptors present on their surface.

Agglutination requires a sufficient ionic strength (usually 0.15 M sodium chloride) to neutralize the negative charges normally present on the surface of bacteria or red blood cells so that the red blood cells may come into contact. Conversely, the addition of a solution of too high ionic strength may provoke agglutination in the absence of antibodies. Sera may not agglutinate at the usual molarity, but the reaction may occur in the presence of an albumin medium. For mixtures of several cell types (e.g., red blood cells and leukocytes), the two cell types separately agglutinate. This observation is compatible with the lattice theory of precipitation. Agglutination is the result of both a decreases in repulsive electrostatic forces between cells (the antibodies acting, then, directly through their positive charges) and of the creation of bridges between cells.

When a bacterium or any other cell is injected into a foreign host, it is lysed, and the immunogens present on the surface and inside the cells stimulate antibody production. However, only antibodies directed against the structures present on the membrane give rise to agglutination reactions. These antigens are called "agglutinogens". It is interesting to note that a given antigen is much more immunogenic as part of the cell surface, as compared to purified extracts, probably because the proximity of other structures on the cell membrane permits development of cooperation phenomena between various B and T lymphocyte populations.

Certain sera lose their ability to agglutinate at high concentrations and must be dilute 100 to 1000-fold to become agglutinating. This is analogous to the prozone phenomenon described earlier in immunoprecipitation. When high concentrations of fluorescein labeled antibodies are incubated with cells, these cells appear to be coated with antibody but do not agglutinate. It is probable that in great antibody excess, simultaneous binding of both antibody sites (one Ig molecule) to two cells is minimized, therefore, agglutination does not occur. This explanation, however, is not entirely adequate, because the presence of nonagglutinating of blocking antibodies may also explain the prozone phenomenon.

Coombs Test (or Antiglobulin Test)

In the 1940s, Coombs developed an ingenious method to detect nonagglutinating antibodies based on the antigenicity of Ig when it is bound to red blood cells by its Fab fragment. The Ig antigenic site (Fc fragments) can then be recognized by anti-Ig sera. The incubation of antibody-coated red blood cells with anti-Ig serum provokes agglutination (indirect Coombs test). A modification used to study hemolytic anemias is the direct Coombs test, which consists of incubating a patient's red blood cells coated in vivo with incomplete antibodies with the antibodies with the anti-Ig serum. A variant of the Coombs test is to use anticomplement antibodies that bind to complement molecules previously fixed onto antibody-coated red blood cells.

Agglutination Applications in Serology

Hemagglutination techniques used in serology are semiquantitative procedures in which variable antiserum dilutions are mixed with a given cell or particle concentration, either in a tube or on a slide.

The mixture is incubated at 4°C for variable times (5 to 60 min) and then agitated. Agglutination is read either visually (macroscopically) or under the microscope. Macroscopic reading detects, for red blood cells, the aspect of the sedimented pellet that is compact and has neat edges (negative reaction) or is more outstretched with irregular edges (positive reaction). The pellet may be resuspended by agitation and the amount of agglutinates noted in the medium. Numerous antibodies agglutinate only at 4°C (so-called cold agglutinins). The agglutination titer is defined as the highest dilution that gives agglutination. This titer is generally evaluated at one dilution and gives only a semiquantitative idea of the amount of antibodies present. A failure to note the presence or absence of agglutination in one tube (one dilution) represents a 100% error in evaluating the quantity of antibody. Titers are influenced not only by the amount of antibody present but also by their affinity and the number and distribution of antigenic determinants present on the indicator cells. This multiplicity of factors explains why agglutination gives only relative information, thus allowing recognition of positive antisera without quantitation. More recently, the introduction of particle counters and autoanalyzers has permitted fairly precise definition of 50% of hemagglutinating doses (HD_{50}).

Passive Hemagglutination

The agglutination target cell may be a red blood cell, a bacterium, or an inert particle (latex, bentonite). Passive agglutination involves precooling of red blood cells and inert particles by soluble antigen. This coupling onto the red blood cell may be accomplished by various means, which most classically consist of treating the red blood cells with tannic acid that makes them capable of reacting spontaneously with soluble antigens, like proteins or nucleic acids. Another method involves mixing red cells and antigens in the presence of chromium chloride or glutaraldehyde. It should be noted that some materials, namely, certain polysaccharides, spontaneously bind to red blood cells. The inhibition of passive hemagglutination by soluble antigen permits antigen quantification. A major application is in the measurement of plasma or urinary hormones including urinary chorionic gonadotropins for the detection of pregnancy. Here, a standard antiserum is incubated with test urine; this incubation is followed by incubation with red blood cells or particles coated with the hormone under study. Agglutination does not occur (positive test) if the urine contained the hormone, because the hormone has previously neutralized the agglutinating antibodies in the antisera.

Neutralization Reactions

The preceding techniques are based on particulate conformation of certain antigens, which permits direct observation of macroscopic agglutination. Macromolecular or soluble cellular antigens may also be studied if endogenous (intrinsic) functions are lost (inhibited) after combination with antibody.

Enzyme Neutralization

An antibody may inhibit the enzymatic activity of a protein if the enzymatic site is identical or close to the antigenic site. Antibody binding then masks the enzymatic site. Enzymatic and antigenic sites however, may be relatively independent. For example, antiperoxidase antibodies do not inhibit the biologic activity of this enzyme. A common application of enzyme neutralization is in the field of bacteriology (detection of antistreptolysins, antistreptokinases and antistaphylolysins).

Toxin Neutralization

Antitoxin antibodies specifically neutralize the biologic effects of toxins. The lethal, inflammatory effects induced by toxins in animals may be neutralize by injecting an antitoxin serum. Toxin neutralization may also be observed in vivo. Thus, human subjects who possess antitoxin antibodies do not show the inflammatory reaction normally seen after intradermal injection of a small amount of toxin. This is the principle of Schick's (for diphtheria) and Dick's (for scarlet fever) reactions.

Virus Neutralization

Antivirus antibodies decrease virus virulence, as expressed by cytopathogenicity. They also inhibit viral growth, and these activities may be used for purposes of quantitation. Additionally, one may also examine the ability of antibody to inhibit viral hemagglutination, which normally occurs when virus is added to red blood cells. This phenomenon is used in Hirst's reaction for influenza.

Phage Neutralization

Phages that infect bacteria have the property of provoking lysis. This characteristic property permits phage enumeration: bacteria are incubated with various phage dilutions, which are then placed into gelose-containing petri dishes. One then counts lysis plates, in which bacteria have been destroyed by phages. If the phage suspension is incubated with an anti phage antiserum before being added to bacteria, plaque (lysis) formation is inhibited. Study of various antiserum dilutions permits evaluation of the minimum quantity of antibody required to produce phage inactivation and thus the testing of antiserum in a sensitive and quantitative way. The technique may be applied to studies of other antigens by fixing them onto phages.

Reactions using Complement

Numerous serologic reactions are based on the ability of CIq, and then of other complement components, to bind to immune complexes. Complement present in the initial serum is inactivated by heating at 56°C for 30 min, thus destroying CIq. Complement is obtained from a standardized source. Fixation of complement is quantitated by measuring the amount of complement consumed (complement fixation) or by studying the biologic effects of complement activation (cytolysis, immunoadherence).

Complement Fixation

An antigenic standard preparation is incubated with the test serum, which was previously heat inactivated. Complement, titillated by hemolytic dosage, is then added to these reagents and the preparation is heated at 37°C for 30 min (or at 4°C for 16 hr). At the end of this fixation phase, sheep red blood cells and a standardized antisheep erythrocyte serum are added. The preparation is again incubated at 37°C for 15 to 30 min, and the amount of complement consumed in the latter reaction is determined. If the test serum contained complement-fixing antibodies directed against the antigen preparation, the complement activity will decrease during the fixation phase, and subsequent sheep red blood cell hemolysis will be proportionately decreased. One must initially verify that the serum in question does not by itself inactivate complement in the absence of antigen. This kind of anticomplementary property is present particularly in certain antisera that contain soluble immune complexes that bind CIq.

A seldom-used variant of complement fixation is conglutination. Conglutinin, present in normal ruminant serum, the property of binding to the hidden determinants of the third component of complement after C3 fixation onto immune complexes. It is a euglobulin of molecular weight 750,000. Conglutinin agglutinates sensitized erythrocytes at subagglutinating antibody concentrations and infrahemolytic complement concentrations. Immunoconglutinin is an antibody that reacts similarly to conglutinin with hidden C3 determinants (perhaps also C4), somewhat analogous to rheumatoid factors (antibodies that react with IgG antigenic determinants). Conglutinin or immunoconglutinin may be used in the complement fixation reaction by hemagglutination utiration before and-after incubation with a mixture of immune complexes and complement.

Cytolysis

Antibodies directed against cellular structures may provoke lysis of the cells that bear these structures after binding the nine complement factors. Cellular lysis is easiest to evaluate for erythrocytes, because hemoglobin release is measured. With nucleated cells, the number of dead cells may be determined by uptake of stain (trypan blue) or by measuring the release of ^{51}Cr previously fixed onto the cells. In this procedure, one incubates cells with different antiserum and complement dilutions for 90 min at 37°C followed immediately by estimation of cellular lysis. Prozone phenomena may be observed. Results are expressed as a hemolysis or cytotoxicity index:

$$\frac{\text{\% of dead cells in the presence of antiserum + complement} - \text{\% of dead cells in the presence of complement alone}}{\text{100\%—dead-cells in the presence of complement alone}} \times 100$$

This phenomenon also explains the treponema immobilization reaction. Treponema in the presence of antitreponema antiserum and complement lose their mobility.

Immunoadherence and Opsonization

Red blood cells and primate platelets, like many other cells, contain surface receptors for the third complement factor (C3). If one adds to particular antigens (bacteria, viruses, leukocytes) their corresponding antibodies and complement, the particles are coated with C3. If admixture with red blood cells or platelets (considered here as "indicators") occurs, the particulate antigens, coated with C3, bind onto red blood cells and platelets and provoke agglutination. This immunoadherence reaction was initially described by Nelson.

Opsonization is closely similar to immunoadherence. Macrophages have surface receptors for the Fc fragment of IgG and for C3. The ability of Macrophages to phagocytose bacteria is strongly augmented by prior incubation of bacteria with specific antibodies and complement. Macrophages may, for example, be obtained from the mouse peritoneal cavity. One may study, in a semiquantitative way, the ability of antibodies to Enhance macrophage phagocytosis of bacteria.

9

ANTIBODIES

HEMOPOIESIS—DEVELOPMENT OF BLOOD CELLS

A Common Stem Cell

The majority of the cell types involved in the immune system are produced from a common *hemopoietic stem cell* (HSC) and develop through the process of differentiation into functionally mature blood cells of all different lineages, e.g. monocytes, platelets, lymphocytes, etc. These stem cells are replicating self-renewing cells, which in early embryonic life are found in the yolk sac and then in the fetal liver, spleen and bone marrow. After birth the bone marrow contains the HSCs. The lineage of cells differentiating from the HSC is determined by the micro-environment of HSC and requires contact with stromal cells and interaction with particular cytokines. These interactions are responsible for switching on specific genes coding for molecules required for the function of the different cell types, e.g. those used for phagocytosis in macrophages and neutrophils, and the receptors on lymphocytes which determine specificity for foreign molecules (antigens). This is, broadly speaking, the process of differentiation.

Stromal Cells

Stromal cells, including epithelial cells and macrophages, are necessary for the differentiation of stem cells to cells of a particular lineage, e.g. lymphocytes, and involves direct contact of the stromal cell with the stem cell. Within the fetal liver, and in the thymus and bone marrow, a variety of stromal cells (including macrophages, endothelial cells, epithelial cells, fibroblasts and adipocytes) each kind

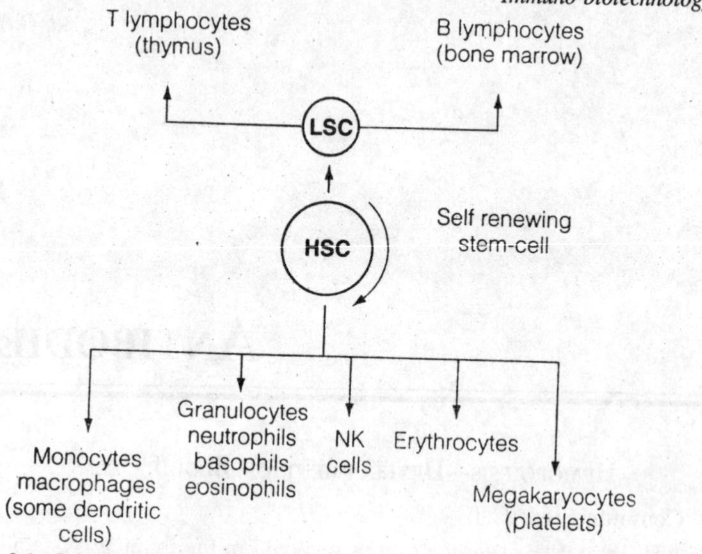

Fig. 9.1. Origin of blood cells (hemopoiesis); LSC—lymphoid stem cell; HSC—hemopoietic stem cell.

of which create discrete foci where different cell types develop. Thus different foci will contain developing granulocytes, monocytes or B cells. Cytokines are essential for this process, and it is thought that adhesion molecules also play an important role.

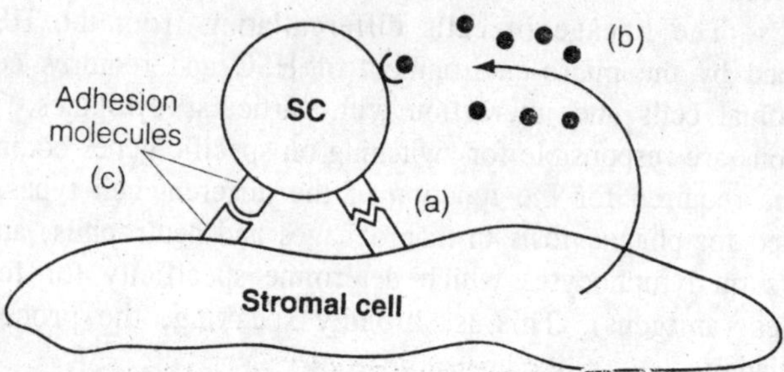

Fig. 9.2. Role of stromal cells in hemopoiesis. (a) Stromal cell bound cytokine (e.g., stem cell factor) and (b) released cytokines (e.g. IL 7) determine the differentiation pathway of the stem cell (SC) attached through (c) specific adhesion molecules (e.g. CD44) on the SC attached to hyaluronic acid molecules on the stromal cell.

The Role of Cytokines

Different cytokines are important for renewal of HSC and their differentiation into the different functionally mature blood cell types. Although an oversimplification, the process related to HSC regenerated

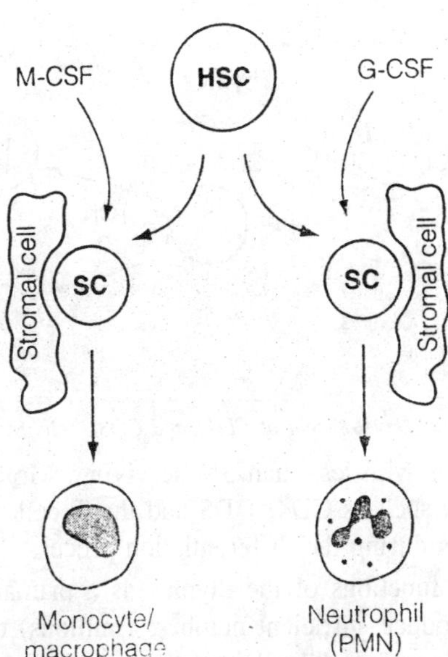

Fig. 9.3. Different cytokines and stromal cells induce different pathways of differentiation.

depend largely on SCE, IL-1 and IL-3, whereas the development of granulocytes and monocytes, for example, involve production of *monocyte colony stimulating factor* (M-CSE) and *granulocyte colony stimulating factor* (G-CSF), along others, by the stromal cells. Interaction of the stem cells with stromal cells and with M-CSE or G-CSF results in the development of monocytes and granulocytes, respectively. Other cytokines are important for the early differentiation of T cells in the thymus and B cells in particular location within the bone marrow.

T Cells are Produced in the Thymus

The thymus is derived from the third and fourth pharyngeal pouches during embryonic life and attracts (by chemoattractive molecules) circulating T cell precursors derived from HSC in the bone marrow. These precursors differentiate into functional T lymphocytes under the influence of thymic stromal cells and cytokines. In the thymus, precursors (now thymocytes) associate with cortical epithelial nurse cells which are important in their development. The thymic cortex is the major site of activity and thymocyte proliferation, with a complete turnover of cells approximately every 72 hours. These thymocytes then move into the medulla, where they undergo further differentiation and selection and finally migrate via the circulation to the secondary lymphoid organs/tissues where they are able to respond to microbial antigens. Most of the thymocytes generated each day in the thymus

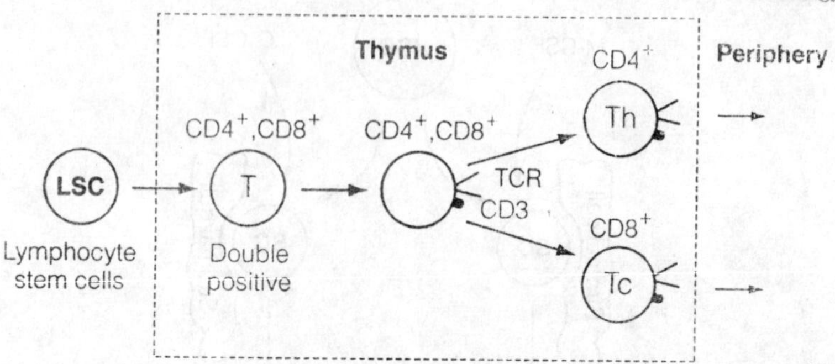

Fig. 9.4. Development of CD4⁺ and CD8⁺ cells in the thymus.

die by apoptosis with less than 5% surviving. Molecules important to T cell function such as CD4, CD8 and the T cell receptor develop at different stages during the differentiation process.

The main functions of the thymus as a primary lymphoid organ are: (a) To produce sufficient numbers (millions) of different T cells each expressing unique T cell receptors such that, within this group, there are at least some cells potentially specific for the huge number of microbial antigens in our environment (generation of diversity); (b) To select for survival those T cells which bind weakly to self MHC molecules (positive selection), but then to eliminate those which bind too strongly to these same self MHC molecules (negative selection) so that the chance for an auto-immune response is minimized. Although dependent on self antigens (MHC molecule), T cell development within the thymus is *independent* of exogenous (foreign) antigens. T cells which survive the selection process migrate to the peripheral lymphoid tissues where they complete their maturation and function to protect against invading microbes. Some extrathymic development of T cells may possibly occur as a result of differentiation of bone marrow precursors in mucosal tissues.

Generation of T Cell Diversity

Each of the very large number of T cells produced in the thymus has only one specificity, coded for by its antigen receptor. Millions of T cells, each with receptors specific for different antigens, are generated by gene rearrangement from multiple (inherited) germ-line genes. The T cell receptor consists of two polypeptide chains, α and β or γ and δ. Each chain is a member of the immunoglobulin superfamily and thus has a uniform domain structure produced by intrachain disulfide bonding. Unlike most proteins produced in the boy, each polypeptide chain of the T cell receptor is coded for by several different genes.

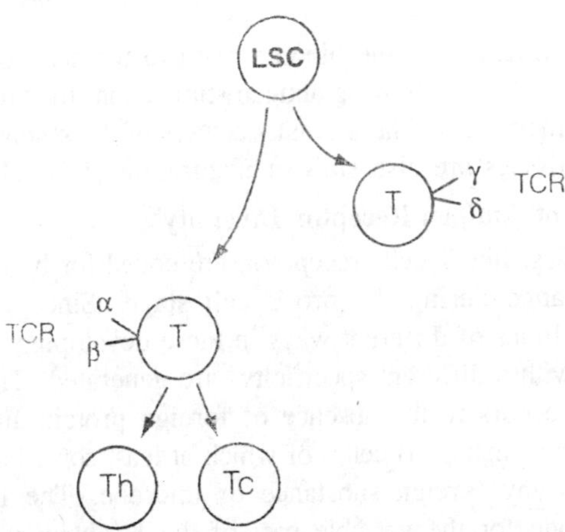

Fig. 9.5. Development of αβ and γδ T cells. Two types of T cells are produced in the thymus with different TCRs (αβ and γδ): the classical T cell (Th and Tc) utilize αβ for their TCR.

Positive and Negative Selection

Once produce in the thymus, T cells undergo selection using their newly produced receptors. T cells with receptors which bind weakly to MHC molecules are selected whilst those with receptors which bind strongly to MHC and self antigens die through apoptosis (central tolerance to self) and are removed by phagocytic macrophages.

B Cells are produced in the Bone Marrow

Birds have a specialized organ in which B cells develop (the Bursa of Fabricius) after which *B cells* were named. However, mammals do not have a discrete specialized organ for B cell development and the microenvironment for HSC differentiation is provided within the fetal liver and after birth, the bone marrow, B cells produce both cell surface and secreted antibodies; the former are their antigen receptors. Antibodies first appear in the cytoplasm of pre-B cells during early B cell differentiation and are then expressed on the cell surface where they function to bind antigen. Two main functions of the bone marrow as a primarily lymphoid organ are to (a) produce large number of B cells, each with unique antigen receptors (antibodies) such that, overall, there is sufficient B cell diversity to recognize the millions of microbial antigens in our environment (Generation of Diversity); (b) eliminate B cell with antigen receptors having high affinity for cell molecules (negative selection: central tolerance).

The early stages of B cell development (like that of T cells) is independent of exogenous antigens. Mature B cells leave the bone marrow and migrate via the blood stream to the secondary lymphoid organs/ tissue where, following antigen stimulation, they become plasma cells and memory cells. The germinal centers of the secondary lymphoid organs and tissues are also sites of maturation of B cells.

Generation of Antigen Receptor Diversity

Antibodies, like T cells receptors, are coded for by multiple genes which rearrange during the pro-B cell stage. Since rearrangement occurs in millions of different ways in these developing cells, many B cells, each with a different specificity, are generated. This generation of diversity occurs in the absence of foreign protein and thus yields large number of mature B cells, of which at least some have specificity for virtually any foreign substance or microbe. The first genes to rearrange code for the variable part of the H chain of the antibody molecule which together with genes of the constant part of the molecule

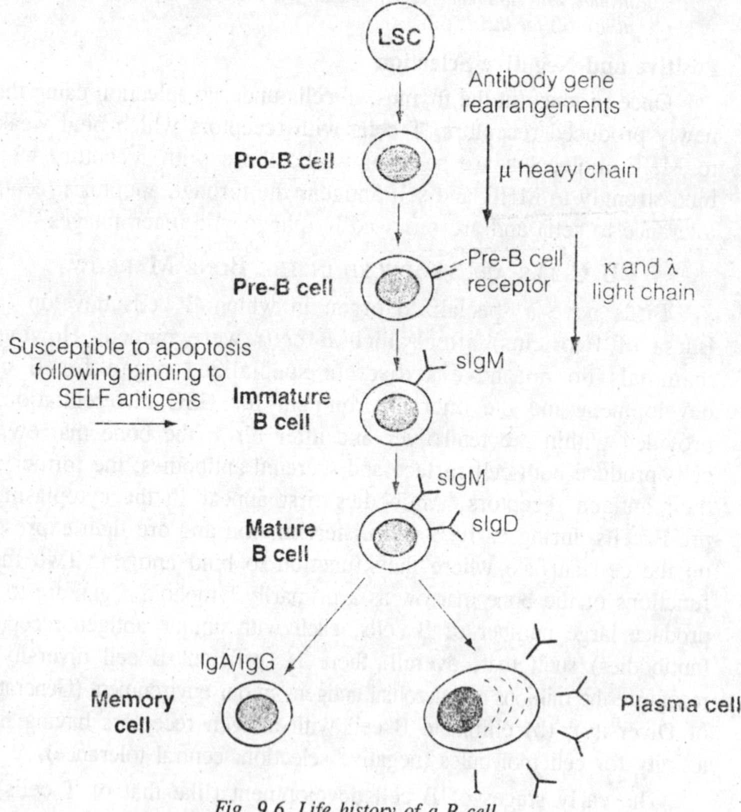

Fig. 9.6. Life history of a B cell.

(and in particular genes which code for the μ H chain) are transcribed first in the differentiation process and appear in the cytoplasm. At this stage, the genes in these Pre-B cells which code for the variable region of the L chains rearrange.

The transcribed H and L chains combine, giving rise to a functional IgM antigen receptor which is then expressed on the surface of the cell (immature B cell). It is during this stage that B cells with high affinity for self antigens are induced to die by apoptosis (negative selection). As in the thymus, the majority of the B cells die during development from production of antigen receptors which cannot be assembled or those directed self antigens.

Development of Immunoglobulin Class Diversity

IgM is the first antibody expressed on a B cell and is initially its antigen receptor. Another class of antibody, IgD, with the same antigen specificity is then expressed on the B cell surface. The variable regions of the IgD and IgM antibodies on the same cell are identical. The variable regions of the IgD and IgM antibodies on the same cell are identical. The function of the IgD is probably that of regulating B cell function when it encounters antigen. Some B cells, in addition to IgM and IgD, may also express other classes of antibody including IgA, IgG or IgE, but all antibodies produced by the same B cell have the same specificity for antigen regardless of their class.

The expression of these other classes of antigen receptor generally requires their activation by antigens and the involvement of T cells and cytokines. The generation of class diversity is achieved through movement (translocation) of variable region genes next to the different

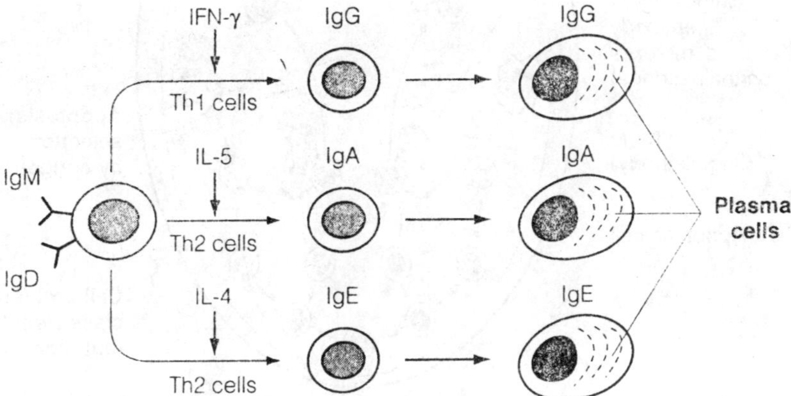

Fig. 9.7. Generation of antibody class diversity.

heavy chain constant region genes. The first antibodies to be produced and found in the fetal circulation and at birth, are IgM antibodies, followed after birth by IgG and IgA antibodies.

Development of other Functional B Cell Molecules

Molecules associated with the B cell receptor complex are expressed early in development to enable assembly of a functional antigen receptor on the B cell surface. Other molecules important for B cell functions, including their ability to present antigen, e.g. MHC class II molecules, also develop early in the life of a B cell.

Germinal Centers as Sites of B Cell Maturation

Germinal centers are unique structures within the secondary lymphoid tissues where two important processes in B cell development occur – the generation of memory cells and the maturation of antibody affinity. Primary B cell follicles in secondary lymphoid tissues, e.g. lymph nodes and spleen, are made up of aggregates of B cells. When B cells in the primary follicle are stimulated by antigen and also receive T cell help, they proliferate, associate with dendritic cells in the follicle (FDC), and being to form the germinal center. These B

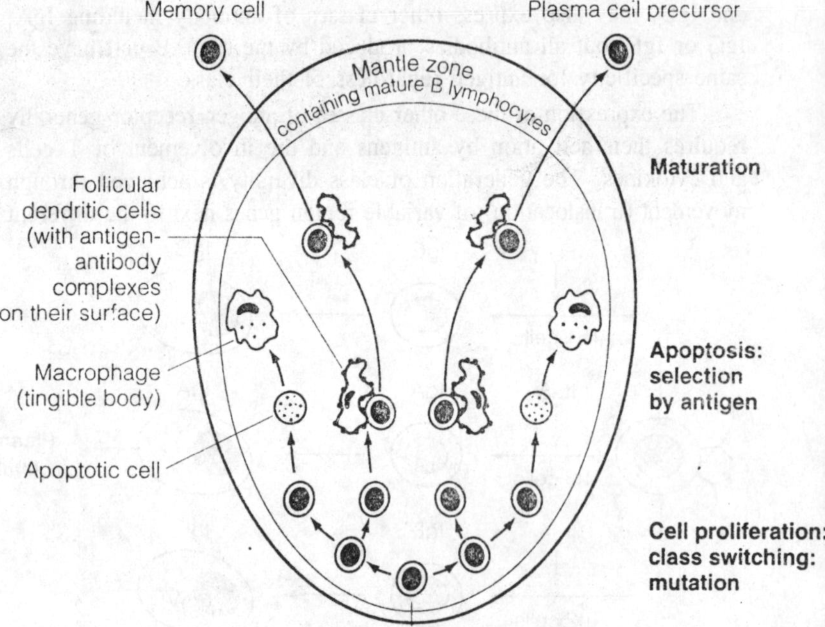

Fig. 9.8. B cell maturation in the germinal center.

cells begin to lose their surface IgM and IgD, and switch to IgG or IgA (usually in mucosal tissues). During this time, there is hypermutation of the variable region genes, and receptors with slightly different amino acid sequences appear on the surface of these B cells.

Many of these modified receptors are unable to bind the same antigen that triggered them and will not be restimulated by this antigen. However, some will be able to bind more strongly to this same antigen, which is often found bound to the surface of the FDC in the form of antibody/antigen complexes. Thus, B cells with higher affinity receptors for the antigen are selected, survive, proliferate and some mature into memory cells which stay in the mantle of the germinal center or join the recirculating lymphocyte pool. Others mature into plasma cells each of which can only synthesize and secrete one class of specific antibody.

Immunity in the Newborn

Lymphocytes in the Newborn

Slightly higher than normal numbers of apparently mature T and B lymphocyte populations (as well as NK cells) are present in the blood of newborn individuals. Even so, the ability to mount an immune response to certain antigens may be lacking. Thus, children under 2 years do not usually make antibody to the polysaccharides of pneumococcus or *Haemophilus influenzae*. In general, the ability to respond to a specific antigen depends on the age at which the individual is exposed to the antigen.

There are a variety of explanations for this sequential appearance of specific immunity, including: (a) Sequential expression of genes coding for receptors for each antigen; (b) Immaturity of B or helper T cell populations; (c) Requirement for further maturation of antigen presenting cells (e.g. macrophages). Since hemophilus polysaccharide conjugated to tetanus toxoid evokes protective anti-polysaccharide antibodies during the first year of life, the neonatal deficiency is likely to be in the Th cell population. Delayed maturation of the $CD4^+$ Th population may contribute to the generally low levels of IgG leading to immunodeficiency in transient hypogammaglobulinemia.

Antibodies in the Newborn

Normally, IgG crosses the placenta (mediated by Fc receptors) and is present in high levels at birth. IgM is produced during fetal development but IgG is not synthesized the novo, until birth. IgA begins to appear in the blood stream at 1–2 months of age. This partly

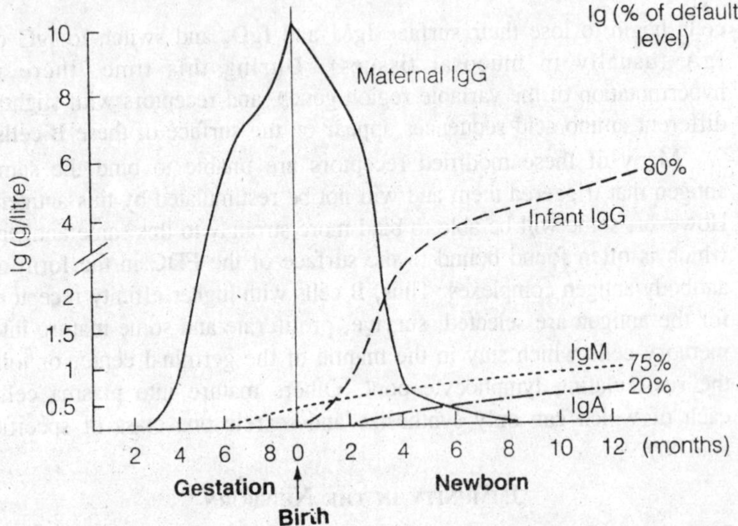

Fig. 9.9. Maternal IgG is actively transported across the placenta and accumulates in the baby's blood until birth.

compensates for the deficiencies in the ability of the infant to initially synthesize antibody through an immature immune system. Furthermore, maternal IgA obtained by the infant from colostrum and milk during nursing coats and infant's gastrointestinal tract and supplies passive mucosal immunity. Unresponsiveness to certain antigens may be related to this passive immunity acquired from the mother since until maternal antibodies are degraded or used up, they may bind antigen and remove it, thereby interfering with development of active immunity.

10

IMMUNO-ELECTROPHORESIS

Immuno-electrophoresis is a method used in both clinical and research laboratories for separating and identifying proteins on the basis of their electrophoretic behaviour and their immunological properties. The introduction of these proteins such as rabbit serum proteins (antigens or immunogens) into another animal such as a goat elicits the production in the host (goat) of other proteins known as antibodies (immunoglobulins). The interaction between the antigen and its antibody is both strong and highly specific. If solutions of antigen and antibody are mixed in different ratios, it is found that at a specific ratio known as the equivalence point, the binding is maximized and a precipitate of antigen-antibody complex is formed. In the clinical laboratory, this technique is used diagnostically. It is utilized in examining certain serum abnormalities especially those involving immunoglobulins, and also semi-quantitative protein analysis of urine, cerebrospinal fluid, pleural fluids, etc. In the research laboratory, this technique may be used to monitor antigen and/or antibody purifications, to detect impurities, to analyze soluble antigens of plant and animal tissues and microbial extracts.

In immuno-electrophoresis, the proteins are first subjected to an electric field and separated by zone electrophoresis on the basis of their different charge-to-mass ratios. The electric field is then shut off and antibodies to the proteins are introduced into the system. Diffusion of both antigen and antibody takes place and at a particular locus, the equivalence point is reached resulting in precipitation. A number of different configurations have been developed to separate, characterize and quantify proteins depending on the emphasis. In classical immunoelectrophoresis, the antibody is added in a trough cut out of the electrophoretic gel after electrophoresis.

This module is to demonstrate the use of immuno-electrophoresis to separate and characterize a mixture of (rabbit serum) proteins as well as to examine the specificity of the antigen-antibody interaction.

Safety Guidelines

Boiling agarose can spatter and cause severe burns. When heating agarose wear safety goggles and hot gloves. Latex gloves should be worn throughout the procedure, especially when handling solutions or gels containing methylene blue. Prior to turning electrophoresis power pack on, be sure that the chamber is level and that the work surface is dry. Wear gloves and safety goggles when operating electrophoresis apparatus.

Experimental Outline

Three hour laboratory with overnight Incubation.

Prepare and pour gel - 20 minutes
Load samples - 10 minutes
Electrophoresis - up to 2 hours
Load samples - 10 minutes
Data analysis - 30 minutes

Prepare and pour agarose gel

Transfer cooled gel to electrophoresis chamber
Cut 2 small holes in gel
Fill chamber with buffer
Load rabbit serum solution
Electrophorese
Cut 2 parallel troughs in gel
Load troughs with anti-rabbit IgG and anti-rabbit whole serum
Allow precipitates to develop

Materials

2 g of agarose
500 ml of TAE (Tris-Acetate-EDTA) buffer, pH 8.0
10 μl of rabbit whole serum
200 μl of goat anti-rabbit serum
10 μl of rabbit albumin
200 μl of goat anti-rabbit IgG
Horizontal electrophoresis apparatus

Pre-lab Preparation

TAE (50X stock, 1 liter)

242 g Tris base

57.1 ml glacial acetic acid

37.2 g $Na_2EDTA \cdot H_2O$

pH 8.5

Dilute antiserum in TAE if necessary

Method

1. Prepare buffer as follows:

   Stock solution- Dissolve 121 g trishydroxymethyl-amino-methane (Tris) and 17 g of ethylenediaminetetracetate, disodium salt (EDTA) in 450 ml of deionized water. Adjust pH to 8.0 with glacial acetic acid (25 ml). Make up to 500 ml with water.

   Working buffer - dilute stock 50 times
2. Prepare agarose gel as follows:

   Mix 0.5 g of agarose and 50 ml of diluted TAE buffer in a 250 ml Erlenmeyer flask. Dissolve by boiling in a microwave oven or on a hot plate. Allow to cool to 50 degrees Celsius. If gel does not stay on slide and runs off edges, clean slides in alcohol. Temperature of molten gel is critical to be at 50 degrees Celsius.
3. Coat a glass plate by applying 2% agar solution with a small brush.
4. Cast the gel on to the glass plate by pouring the molten gel to a thickness of 1-2 mm and allow to cool.
5. Cut 2 or more small holes in the gel by punching the agarose gel with a pasteur pipette connected to a water aspirator.
6. Transfer the gel to the electrophoresis apparatus and add enough buffer to just cover the gel.
7. Add 2 drops of a bromophenol blue solution to the rabbit albumin and the rabbit whole serum solutions.
8. Load rabbit albumin into left well and rabbit whole serum into the right well with a micropipette or capillary tubing.
9. Elevate slide 2-3 mm above electrophoresis buffer solution. Make contact between the buffer and agarose by using a wick of filter paper presoaked in electrophoresis buffer. One end of the wick should overlap the end of the gel by 3-4 mm and the other end submerged in electrophoresis buffer.
10. Electrophorese at 50 V and 30 mA for 1 hour

11. Carefully remove the gel from the electrophoresis chamber. Cut two troughs, one between the two wells and the other to the right of the whole serum well. Use a sharp blade to delineate the troughs and a Pasteur pipette tip to remove the cut gel.
12. Load the left trough with goat anti-rabbit whole serum and the right trough with goat anti-rabbit IgG.
13. Place the gel in a closed humidifying chamber containing moistened paper towels and allow diffusion to take place over a 24 hour period or until a visible precipitate is formed.

Results

1. Note the formation of arcs of precipitate in the gel.
2. Identify the number of proteins in the rabbit whole serum from the number of arcs of precipitate.
3. Identify albumin and IgG in the whole rabbit serum from comparison with the pure albumin and the pure anti-IgG segments.

The actual number of arcs of precipitate will depend on the source of your rabbit serum and the efficiency of separation.

11

IMMUNO-BIOTECHNOLOGY OF YEAST CELLS

The phenomenon of antibody binding to insolubilized antigen as part of a cell membrane or cell wall may lead to cross-linking of the cells called *agglutination*. Agglutination occurs as long as there is an adequate concentration of antigen on the cell surface and the cross-linking protein is at least divalent. The affinity constant is the quantitative measure of the strength of binding of the agglutinating protein. It must be great enough to maintain an equilibrium in order for two or more cells to be bound together by the antibody.

Antibody agglutination of cells is a major way in which the immune system functions to clear the body of the antigen-containing cell. Agglutination in vitro is a valuable diagnostic and research tool to give one a first approximation of the presence or absence of an antigen. Besides antibody there is a class of naturally occurring carbohydrate-binding proteins which will agglutinate cells. These proteins, typically found in plant seeds and animal tissues, are called lectins. They are defined as proteins of non-immunologic origin, having a carbohydrate-binding specificity, and lacking an apparent enzyme activity.

The use of a lectin as an agglutinating protein exhibits their function in the living organism. In leguminous plants this binding may be important to capture microorganisms that will be enclosed in nodules. Other plant lectins may bind soil microorganisms that are potential pathogens—a rudimentary plant immune system. Concanavalin A, a lectin isolated from Jack bean meal, has a carbohydrate-binding specificity for alpha-linked mannose and glucose residues. The specificity of the lectin combining site is determined by observing the degree of agglutination inhibition by using a bank of mono- and disaccharide inhibitors. Antibody combining site specificity can be determined using a similar inhibition method. The inhibiting compounds for carbohydrate-specific antibody are generally more complex in size, makeup, and conformation due to the larger antibody combining site relative to that of most lectin combining sites.

Yeast cells, *Saccharomyces cerevisiae*, have a cell wall component called mannan that is a polymer of alpha-linked mannose residues. Concanavalin A will bind to mannan and with an adequate concentration of cells agglutinate the yeast cells. There is no biological importance to this phenomenon. However, it does exhibit agglutination quite well.

Antibodies and lectins are proteins with combining sites of defined specificity. The nature of the combining site specificity is determined by studying the size, shape, and chemical groupings within the combining site. The interaction of chemical groupings of a ligand and a combining site is required to maintain a "fit" or association which stabilizes the complex. Antibodies especially to protein and polysaccharide antigens have a combining site specificity which when probed using polypeptides (short chains of amino acids) or oligosaccharides (short chains of carbohydrates) conforms to a site of variable depth and conformation. Molecules with the proper shape and chemical grouping to fill the combining site will inhibit the binding of other molecules with a like specificity for the combining site. Lectins do not have combining sites as large and complex as carbohydrate-binding antibodies.

An approximation of antibody and lectin combining site specificity can be made by studying a battery of molecules of known chemical makeup and structure. Lectin combining site specificity can be investigated using agglutination inhibition. The inhibiting molecules are soluble. The molecule to which the lectin is being inhibited from binding is that which is on the cell surface. The competition for the combining site between the insolubilized cell surface component and the solubilized inhibitor is concentration dependent and also dependent on complementarity of fit as described above. The more complementary an inhibitor is to the combining site the more effectively it will compete in binding to the combining site. Complementarity includes, in part, the chemical interaction that occurs between an inhibitor and a combining site. Two molecules are competing for the combining site in agglutination inhibition. The insolubilized molecule on the yeast cell is called mannan, a yeast cell wall protein-polysaccharide complex where the polysaccharide is mainly alpha-linked mannose chains; and the inhibitor is a soluble inhibitor molecule of a mono or disaccharide

Safety Guidelines

Safety - cleanup, waste disposal

Yeast - keep away from cell culture area to prevent contamination.

Autoclave all tubes containing yeast

Con A solutions can be disposed of in sink drain

PBS can be disposed in sink drain.

Experimental Outline

Timetable of events

Pipetting - 20 minutes

Ice incubation - 10 minutes

Macroscopic observation - 5 minutes

Wet mounts and microscopy - 5 minutes

Counting Cells/agglutinate - 40 minutes

Histogram - 20 minutes

Data Analysis - 20 minutes

Procedures

1. Label test tubes
2. Pipet control and lectin solutions
3. Pipet inhibitor solutions
4. Pipet yeast suspension into control and lectin solutions
5. Ice bath for ten minutes
6. Observe for agglutination macroscopically and microscopically

Materials

Yeast cell suspension (Fleischmann's dry yeast)

Phosphate buffered saline (PBS)

Concanavalin A, Sigma Chemical and dilutions (in PBS) of 1 mg ml, and 0.1 mg/ml

.1 M Lactose, 0.5M D-Galactose, 0.5M D-Glucose, 0.5M alphamethyl D-mannoside, 0.5M Sucrose

Pasteur pipettes

1, 5, and 10 milliliter pipettes

15 mm diameter test tubes

Microscope slides, cover slips

Microscope

Hand lens

Pre-lab Preparation

PBS: PBS can be purchased as a 1x or 10x solution, or as a powder. If prepared in the laboratory, dissolve the following in 150 ml distilled water:

2.19 g NaCI

1.28 g KH_2PO_4

2.63 g $Na_2HPO_4 \cdot 7H_2O$

Adjust pH to 7.2, add distilled water to bring volume to 250 ml

Yeast suspension - [Resuspend 600 mg of Fleishmann's dry yeast in 50 ml PBS (need 10 ml suspension for whole module)] 10 minutes (Can be stored in refrigerator for 2 days)

Con A dilution from lyophilized powder - 20 minutes (can be stored In refrigerator for days if solution is kept sterile or for weeks if stored frozen at -20°C)

Con A dilution from concentrated solution - 10 minutes (can be stored in refrigerator for days if solution is kept sterile or for weeks if stored frozen at -20°C)

Method

Agglutination

1. Label four 15 mm tubes:

   Tube A: lectin 500

   Tube B: lectin 100

   Tube C: lectin 10

   Tube D: PBS control
2. Using a 1 ml pipette, pipet 0.5 ml of 1 mg/ml Con A Into Tube A.
3. Using a 1 ml pipette, pipet 0.1 ml of 1 mg/ml Con A, and 0.4 ml of PBS into Tube B. Mix.
4. Using a 1 ml pipette, pipet 0.1 ml of 0.1 mg/ml Con A, and 0.4 ml of PBS into Tube C. Mix.
5. Using a 1 ml pipette, pipet 0.5 ml of PBS into Tube D.
6. Using a 5 ml pipette, pipet 0.5 ml of stock culture of yeast cells into each tube - tubes A, B, C, and D. Mix.
7. Place the tubes on ice for 10 minutes, mix occasionally - every three or four minutes.

8. Mix and observe for agglutination.
9. Mix and prepare a wet mount by placing one drop of the suspension on a microscope slide and covering the drop with a 22 × 22 mm coverslip.
10. Place the slide and wet mount on the stage of a light microscope. Find the field of cells beneath the cover slip by using the 10X objective. Then view the cells by light microscopy at 400x magnification.

Agglutination inhibition

1. Label seven 15 × 150 mm tubes:

   Tube A: lectin 500
   Tube B: Lactose
   Tube C: Galactose
   Tube D: Glucose
   Tube E: alpha-methyl mannoside
   Tube F: Sucrose
   Tube G: PBS control
2. Pipet 0.5 ml of 2 mg/ml Con A into each tube *except* Tube G. Page 134
3. Pipet 0.5 ml PBS into Tube A, and 1.0 ml PBS into Tube G.
4. Pipet 0.5 ml carbohydrate inhibitor into Tubes B, C, D, E, and F. Mix.
5. Pipette 1.0 ml of stock culture of yeast cells into each tube. Mix.
6. Place the tubes onto ice for 10 minutes, mix occasionally every three or four minutes.
7. Mix and observe for agglutination
8. Mix the cells, prepare a wet mount and view cells by light microscopy at 100x magnification.

Results

For observing agglutination macroscopically use a hand lens to observe the suspension of yeast cells. Score each tube using - to ++++ system with the PBS control (Tube D) being the negative (-) score control.

For microscopic quantitative analysis randomly select fields of cells on the microscope slide. The objective is to view random fields of cells and count the single cells and number of cells in agglutinates. So as not to bias results by looking for a "good field with lots of cells" randomize the selection of fields. Random selection is best accomplished by moving the microscope stage vertically or horizontally without viewing the movement through the ocular(s). Move the stage slightly for high powered fields since a small move changes the field completely. The goal is to gather a sufficient quantity of data to graph results so that a conclusion(s) can be made. Ideally every cell and/or agglutinate on the slide should be scored. This is impractical. Arbitrarily choose to count cells and agglutinates in ten high powered fields (hpf). Single cells are counted as one, agglutinates are categorized in the following way 2-3 cells/agglutinate, 5-9 cells/agglutinate, 10-14 cells/agglutinate, 15-20 cells/agglutinate, and 20+ cells/agglutinate. The numbers of cells in an agglutinate can be counted by focusing in and out of the agglutinate using your fine focus adjustment. Cells above and below the plane of focus will come into view and can be scored or counted. If an agglutinate obviously has more than 20 cells don't bother to count the cells rather score it as a 20+.

Control tube

Scan the Control tube D wet mount using the 4X or 10X objective for a field with a significant number of cells. Adjust the microscope to a higher power of magnification, preferably 400X and observe the cells. Count the number of free cells and cells in agglutinates in this first high powered field (hpf). Move the stage horizontally and vertically and after each move count the number of free cells and cells in agglutinates. Count 10 hpf. Make a table indicating the number of cells in a clump and the number of free cells. This Control tube data is your background nonspecific agglutination data.

Experimental tube

Prepare a wet mount of experimental tubes A - C. Agglutinated cells will appear as clumps of cells in a "cluster of grapes" appearance. Scan for a field with a significant number of cells. Adjust the microscope to a higher power setting, preferably 400X and observe the cells. Arbitrarily pick a field with a significant number of cells. Count the number of free cells and cells in agglutinates in each tube. Count agglutinates and/or free cells in 10 hpf. Complete the table by indicating the number of cells in a clump and the number of free cells.

12

IMMUNO-DIFFUSION

Radial immuno-diffusion is a technique that can quantitatively determine the level of an unknown antigen. Preformed antibody is incorporated into liquid agar which is poured into a Petri dish and allowed to gel. Small wells cut into the agar dish are filled with solutions of known concentration antigen and an unknown solution sample. The antigen solution diffuses outwards from the well in a circular pattern. In the endpoint or Mancini method used in this module, antibody is present in excess amount and diffusion continues until a stable ring of antigen-antibody precipitate forms, generally within 24 to 48 hours.

For each standard antigen concentration an endpoint precipitation ring of a certain diameter is formed. The diameter squared measurements of the rings plotted against the known concentration of each starting standard yields a straight line. From this linear calibration curve the concentration of the unknown antigen sample may be determined. Unlike many gel and liquid precipitation techniques which qualitatively detect antigen, RID is a sensitive quantitative technique that is often used clinically to detect patient levels of blood proteins.

Safety Guidelines

No mouth pipetting.

Students will wear safety glasses.

Dispose of used RID plates through proper laboratory waste disposal procedures.

Experimental Outline

Most kits are designed to be performed quickly, and generally take 10-30 minutes to complete. Incubations to allow for diffusion and ring formation generally occurs within 24 to 48 hours. This lab can be included in another related lab either to illustrate applications of antibodies to start the lab or during an incubation or other waiting period.

Perform test: follow test instructions - generally 10-30 min.

Analyze results: interpret test results by measuring ring diameters about 15 min.

Materials

RID plates

Set of reference standard solutions for antigen

Calibrated ruler, mm.

Pipette gun, 5 microliter or plastic micropipette

Graph paper

Pre-lab Preparation

Generally none, other than obtaining kits and samples.

Method

1. Perform the test using the samples provided, following the directions provided with the kit.
   (a) Add 5 microliters of antigen solution standards to each of four wells. Rinse micropipettes or change tips.
   (b) To a fifth well add 5 microliters of the unknown antigen solution. Snap the cover of the Petri dish back on. Place the dish into a zip lock bag and seal.
   (c) Allow the dish to sit undisturbed for 24-48 hours at room temperature on a flat surface.
   (d) After precipitin rings have reached equilibrium, remove the plate from the bag and place on a light box or back lit plate.
2. Record the results of the test.
   (a) Measure the diameter of each precipitin ring with the ruler to the nearest tenth of a millimeter.
   (b) Square the ring diameters and plot these values on the Y axis of the graph paper versus the known starting antigen concentrations.
   (c) Draw the best fit line through the standard points.
   (d) Using the standard line determine the value of the unknown antigen.

Results

1. Interpret the results of the test. considering any control sample run.
2. Plot the values of precipitin ring diameter squared versus known antigen concentrations. Draw the best fit line through these points. Calculate the value of the unknown antigen concentration from this line.

13

Spleen Cell Preparation

The preparation of lymphocytes from mouse spleen is a technique which readily provides a source of functional T cells, B cells, and macrophages. Unlike the thymus and bone marrow which contain undifferentiated lymphoid and myeloid cells, the spleen is classified as a secondary lymphoid organ along with the lymph nodes and the peripheral blood and contains active differentiated cells. The spleen is easily dissociated into a single cell suspension and provides a high yield (approximately 1-2 $\times$ 10^8 cells per spleen).

Cells obtained from the spleen are an excellent starting material for primary cell cultures. Cells obtained from the spleen may be fractionated into T lymphocytes, B lymphocytes, and macrophages and their activities studied using in vitro systems. Historically this approach along with controlled reconstitutions after fractionation led to discovery of the nature of T cell, B cell, and macrophage cooperation.

After preparing and counting splenic lymphocytes an investigator may proceed with structural or functional studies of the cells. Splenic lymphocytes from immunized animals are key starting cells for the production of hybridomas and monoclonal antibodies. The activity of T lymphocytes sensitized in culture to react against tumor cells is being studied to determine more effective treatments against specific types of cancer.

Instructors should obtain clearance from the animal use committee of their institution for demonstrating that they are complying with accepted federal laboratory procedure guidelines.

If mice are unavailable bovine spleen may be obtained from a slaughterhouse. The fresh tissue should be placed in BSS on ice. Some lymphoid cell lines are available from American Type Culture Collection (ATCC).

Safety Guidelines

1. No mouth pipetting.
2. No ethanol beakers near open flames.
3. Students will wear safety glasses.

Experimental Outline

Sacrifice mice
Remove spleen
Sieve spleen cells through wire mesh
Wash cells
Viable cell count with trypan blue
Cell count with hemocytometer

Materials

Balanced Salt Solution (BSS)
Acetic Acid Counting Solution
Trypan Blue Vital Dye Solution
Beaker Containing 70% Ethanol
Squeeze Bottle Containing 70% Ethanol
Small Surgical Scissors
Large Surgical Scissors
Curved Hemostat
Curved forceps
Rubber Policeman
Stainless Steel Wire Mesh
Glass Centrifuge Tube, 30 ml.
Glass Tubes, 5 ml.
Glass Petri Dish
Stainless Steel Wire Mesh
Hemocytometer
Pasteur Pipettes With Rubber Bulb

Pre-lab Preparation

1. Make up 10X stocks of Balanced Salt Solution (BSS)

Stock #1	Dextrose	10 gm
	KH_2PO_4	0.6 gm
	0.5% phenol red solution	20 ml

Dissolve and bring up to 1000 ml with distilled water

Stock #2	$CaCl_2 \cdot H_2O$	1.86 gm
	KCl	4.0 gm
	NaCl	80.0 gm
	$MgCl_2$	1.04 gm
	$MgSO_4 \cdot 7H_2O$	2.0 gm

Dissolve and bring up to 1000 ml with distilled water.

Mix 10 ml of stock #1 and 10 ml of stock #2, bring up to 100 ml with distilled water. The pH should be 7.2- 7.4.

2. Sterilize all instruments and wire mesh by storing in 70% ethanol. Place in a loosely covered sterile Petri dish and allow to air dry before use. Pasteur pipettes and glass tubes may be purchased sterilized. Alternatively, all materials can be wrapped in aluminum foil and autoclaved (check any plastic ware for stability under heating). Standard microbiology flaming technique can also be used for Instruments.
3. Make up the following solutions for cell counting: 0.5% acetic acid in distilled water (v/v) for lysing red blood cells trypan blue 10x stock, 1 % (w/v) in distilled water
4. Sacrifice mice just before the laboratory in a closed chamber containing 30% CO_2. An alternative, but less-preferred method, is to use cervical dislocation.

Method

1. After the mouse skin has been wet with 70% ethanol, place the left side facing up on the paper towels.
2. Make a cut through the loose skin by pulling gently upwards and using the blunt scissors on the skin flap. This will expose the peritoneal wall.
3. Pull gently in opposite directions on the two sides of the skin incision to expose a wider area of the peritoneal wall.
4. With the small surgical scissors make a cut over the spleen to expose it through the peritoneum.
5. Grasp the spleen with the forceps. Pull up gently and cut away attached connective tissue which appears white using the small surgical scissors.
6. Grasp a folded corner of the stainless steel wire mesh with the hemostat. Insert the mesh into the Petri dish and add about 10 ml of BSS.

7. Put the removed spleen on the wire mesh. With a rubber policeman, rub the spleen back and forth over the mesh. Continue until all the spleen dissociates leaving only white connective tissue on the mesh surface.
8. With a sterile Pasteur pipette bring the cell suspension up and down several times to dissociate large cell clumps. Pipette the cell suspension into the glass centrifuge tube.
9. Allow the cells to settle for 5 minutes to remove larger debris. Use the Pasteur pipette to transfer roughly 2 ml volumes of cell suspensions to 5 ml glass tubes.
10. Centrifuge at 200 x g for 10 minutes. Pour off the supernatant.
11. Resuspend the cell pellet in 2 ml of BSS by vortex mixing or up and down action in a Pasteur pipette.
12. Add .1 ml of trypan blue solution to 1.8 ml of acetic acid solution. Add.1 ml of the cell suspension and mix thoroughly.
13. Fill a hemocytometer chamber with a drop of your mixture and proceed as if doing a white blood cell count. Count both total cells and the fraction which are dead (those that have taken up the blue dye).
14. Remember that a dilution factor of 20 and a volume factor of 10,000 need to be multiplied by the average count of a 4 by 4 square grid. For example, 50 cells in a 4 by 4 grid would yield a cell count of 10 million cells/ml.

Results

Record your grid counts. Average four separate grids. Record the number and per cent of dead cells per grid and average these trypan blue positive counts.

14

ENZYME IMMUNO-FILTRATION

Monoclonal Antibody (MoAB) technology represents a major breakthrough in the field of immunology in that it gives investigators an opportunity to study virtually any immunogenic molecule. As investigators today are still describing new applications of this technology, it is doubtful that George Kohler and Caesar Milstein fully grasped the impact of their seminal 1975 publication on the scientific community. This publication, describing the immortalization of hybrid cells producing antibodies to red blood cells, set the precedent for future publications dealing with fusion parameters enhancement of antibody production, the phenotypic dissection of cell subtypes and new applications and assays utilizing MoAb technology.

Due to the rapid growth of Ig-secreting hybridomas, questions pertaining to the type, amount, and specificity of those MoAbs must be answered quickly to avoid constant culturing. The study of antigen-antibody interactions over the years has resulted in an overabundance of assays capable of analyzing these reactions and suggests that no single method satisfies the requirements of the majority of investigators. As some investigators in the MoAb field screen anywhere from hundreds of wells a week to thousands a day there is an obvious need to employ rapid, simple, and sensitive assays. The development of the enzyme-linked immunoassays (EIA) offers many distinct advantages over other conventional antigen-antibody reactions, such as radio-immunoassays (RIA). In addition to the elimination of costly and hazardous radiolabled compounds, EIA's may be read visually as well as quantitatively and eliminate the need for time consuming gamma counting.

This chapter discusses the utility and versatility of the enzyme immuno-filtration assays (EIFA) used in our laboratory to determine

MoAb class, concentration and specificity. Immuno-filtration was first described by P. H. Cleveland in 1979 when he utilized an immuno-filtration device to immobilize antigens. Enzyme-conjugated antisera were first applied to immuno-filtration by Richman et al. when they analyzed antigens on herpes viruses. Subsequently, MoAb technology was introduced into the EIFA to exploit the specificity and precision of these antibodies and the ease in which an EIFA can be performed. While reading this review one should understand that most of the concepts discussed here are applicable to both the EIA and RIA. The EIFA we describe here is well suited for the screening of both mouse and human hybridoma supernatants in a rapid, simple and sensitive manner for the detection of MoAbs to either soluble, particulate or cellular antigens.

Principles of Immuno-filtration

In concept and principle the EIFA is elegantly simple and nearly identical to conventional EIA's and RIA's. Each of these assays exploits the physical nature of antigen-antibody interactions and are performed in either direct or indirect manners. In the direct EIFA an enzyme is covalently conjugated to the MoAb itself for detecting reactivity to its antigen whereas in the indirect EIFA an enzyme-conjugated Ig class specific antiglobulin is use to react with the test MoAb already bound to its antigen. In addition, modifications on the indirect assay method allows for amplification of the signal (colour) to noise (background) ratio. The quantification of this signal to noise ratio is the strength of any immunoassay, whether it be by quantifiable colour change or the amount of radioactivity bound. Enzyme assays may be preferred de to the longer shelf life of reagents, the elimination of radioactive waste, and comparable scientific results. In addition, a variety of different enzymes and both fluorogenic and chromogenic substrates may be used,

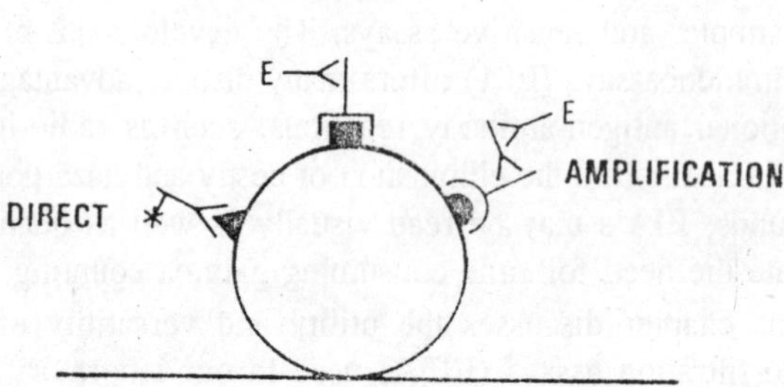

Fig 14.1. Option in immunoassay protocols with monoclonal antibodies.

each of which may have special utility to any specific investigational needs. EIFA has the added feature in that it utilizes the principle of liquid surface tension to provide an incubation chamber that acts both as a solid phase reaction site and a filtration device. EIFA's do this by utilizing a 96-well microtiter plate modified with a small hole in the bottom of each well. Over each hole is placed a filter which also acts as the reaction surface. These filters are available in either glass fibre, cotton lentil, nitrocellulose or polycarbonate form. Each type of filter differs from the others in surface area, flow rate and amount of nonspecifically bound protein; all types serve as a solid phase incubation chamber until a vacuum is applied uniformly across the bottom of the plate. Application of a protein or TWEEN 20 supplemented phosphate buffered saline (gelatin buffer) to each well completes the vacuum facilitated wash step.

Methodology in Enzyme Immuno-filtration

Assay Procedure

The EIFA's are utilized in our lab to identify the type, concentration, and specificity of MoAbs. This occurs in modified microtiter plates available through V & P Scientific, Inc. or Millipore. The bottom of each well contains a 0.6 um hole covered by a filter which allows the passage of liquids, upon application of a vacuum, while immobilizing particulate material. Both soluble and insoluble antigens can be immobilized onto the surface of the filter by vacuum drying.

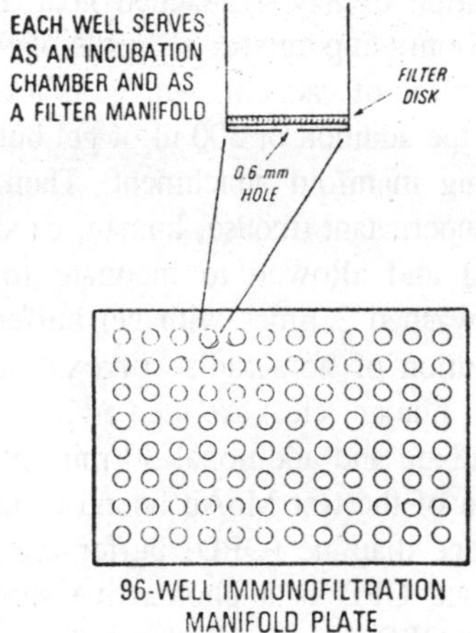

Fig 14.2. Schematic of the 96-well immuno-filtration manifold plate.

After the filters have been absorbed with antigen they are washed by application of 380 mm/Hg of vacuum to all 96 wells at once. In this way, 0.3 ml of fluid can be drawn through the 0.6 mm hole in each of the 96 filters within 5 sec, thus efficiently removing soluble material and retaining particulate matter on the filter discs. A primary antibody or labled reagent in then incubated with the antigen under study, then washed again by vacuum. Three thorough washes of each 96-well plate can be completed within 45 sec. The enzyme-conjugated antibody is added, allowed to incubate, washed, then developed with substrate. When read visually, this EIA can be completed in under two hours. In many cases the colour change reaction by the enzyme can be quantitated objectively on EIA readers.

The EIA procedures we employ are divided into two main approaches. The first is the detection of Ig class and concentration while the second is the analysis of MoAb specificity.

For Ig class and concentration, 50 ul of an appropriate dilution of a goat IgG anti-mouse (or human) polyvalent Ig antisera in PBS or carbonate buffer are added to each well. For Ig class determination, 50 ul of an affinity purified goat IgG anti-heavy chain (mouse or human) specific (mu, gamma, etc.) antisera, used at an appropriate dilution, are then added to the wells and allowed to incubate at room temperature for 30 min. A vacuum (380 mm/Hg) is then applied to the filter manifold for 10 min to dry the antibody to the filter material. Then, 50 ul of fetal calf serum (FCS) buffer are added to each well to allow for the resolubilization of any unattached goat antisera. After a 15 min incubation at room temperature, all wells of the 96-well plate are evacuated by application of vacuum. All filters are then washed under vacuum 3 times by the addition of 300 ul of gel buffer with a repeating syringe and 8 prong manifold attachment. Then, 50ul of each test MoAb hybridoma supernatant (mouse, human, or any other appropriate species) are added and allowed to incubate for 30 min at room temperature, then washed 3 times with gel buffer. Finally, 50 ul of the appropriate dilution of horseradish peroxidase (HRP)-conjugated goat anti-Ig (mouse, human, etc.) are then added to the specified wells and allowed to incubate and additional 30 min. at room temperature. For visual detection of the test MoAb bound to its target antigen, 50 ul of ortho-phenylene diamine (OPD) buffer are added to each well and developed. Since OPD is a photoactive substance it is best to develop the enzyme-OPD reaction in the dark (essentially this entails keeping the 96-well immuno-filtration manifold plate in a drawer) and glancing at it periodically to check for reaction progress.

For MoAb concentration determinations a titration of Ig (e.g., IgG or IgM) of known concentrations is added to some wells in lieu of the test MoAb and developed as described above to generate a standard curve. The test MoAb concentrations are then obtained from this curve. To quantitate the MoAb reactions, each immuno-filtration manifold plate is first sealed to prevent leakage and then 250 ul of OPD buffer is added to each well and incubated in the dark. At the completion of the reaction, 150 ul from each well is transferred to a micro-titer plate containing 50 ul/well of 2.5M H_2SO_4 and subsequently read on a micro-EIA reader. Many of the commercially available EIA readers are equipped for interfacing with a computer to perform whatever kind of data analysis the investigator requires.

Once the mouse or human MoAb secreting hybridomas have been identified it now becomes important to establish the specificity of the test MoAbs. Antigens amenable to EIFA are of two major types: either soluble or particulate. In either case, the antigen is immobilized onto the filters of the immuno-filtration manifold plates, then incubated with the test MoAb, washed, incubated with the appropriate HRP-conjugated antisera, washed, and then developed with OPD buffer and either read visually or quantitated by an EIA reader. In the case where the antigen is a cell surface macromolecule, then log-phase cells are collected, washed in PBS, and resuspended to 4 $\times$ 10^6/ml. A 50 ul suspension of the cells is then added to each well of the immuno-filtration manifold plate followed by the appropriate EIA procedure outlined above.

Buffers

In this format, investigators wishing to adopt our EIA procedures may make the necessary buffers by mixing the specified amounts of indicated reagents and adjusting each to its appropriate pH.

The carbonate buffer is used to attach an antibody to its solid phase support. Gelatin buffer is used for washing. Nearly all incubations and secondary reagent dilutions are made with fetal calf serum buffer. OPD, diluted in citrate buffer with hydrogen peroxide, is the substrate we use for the horseradish peroxidase enzyme.

Conventional Antibody Reagents

All of the conventional reagents we have used in our EIFA procedures are commercially available. We have adapted the EIFA for the qualitative and quantitative analysis of both murine and human MoAbs. Depending upon which type of MoAb is of interest to the investigator their choice of reagents will reflect this. We use affinity purified goat anti-mouse or affinity purified goat anti-human class specific (i.e., IgA, IgD, IgE, IgG, or IgM) antisera. Practically, the IgG and IgM reagents are more useful than the other isotypes. In addition, most of these antibody reagents are available conjugated to an enzyme, such as horseradish peroxidase (HRP) or alkalin phosphatase, or use in the EIA. The data described in this review were obtained with HRP-conjugated reagents. Also, reagents to the kappa and lambda light chains are available. For a detailed list of the commercially available antibody reagents the investigator is encouraged to consult Linscott's Directory of Antibody Reagents. This directory lists virtually every type of reagent available for EIA's and their commercial source.

Since there are so many commercially available sources marketing these reagents, and the lot to lot variability, the investigator should titrate each batch of reagent to optimize for his particular assay.

Cell Lines

The following human cell lines were used to obtain the data described in this review. B lymphoblastoid cells WIL-2, UC 729-6, and 8392; T lymphoblastoid cells CEM, Molt-4, 8402 (the autologous counterpart to the Epstein-Barr virus transformed B cell, 8392), and HPB-ALL; erythroleukemia K562; myelogenous leukemia KG-1; squamous cell carcinomas of the cervix CaSki and Hela; adenocarcinoma of the colon T84; carcinoma of the larynx Hep-2; adenocarcinoma of the lung T291; oat cell carcinomas of the lung NCI-69 and T293; squamous cell carcinomas of the lung Calu-1, SK-MES-1, and T222; melanomas M21 and SK-MEL-28; prostate carcinomas DU 145, H494, Ln-Cap, and PC-3; signet ring cell carcinoma of the stomach Kato-III; carcinoma of the vulva A431; and normal fibroblasts 350Q and WI-38. All cell lines used here were grown in RPMI-1640 media supplemented with 10% fetal calf serum and 2mM L-glutamine at 37°C in a 5% CO_2 humidified incubator.

In addition, red blood cell types A^+, B^-, and O^+, plus purified peripheral blood lymphocytes were analyzed by EIFA.

Cell Preparation for EIFA

Cells in log-phase were washed twice (500 × g/wash) in phosphate-buffered saline (PBS) to remove cytophilic proteins which were either endogenous or from culture serum and resuspended to 4×10^6 target cells/ml in fetal calf serum buffer.

Monoclonal Antibodies

We have used these two types of MoAbs to demonstrate the utility of the EIFA as applied to MoAb technology. Further details pertaining to the production and generation of these MoAbs can be found in the appropriate references. The investigator should note that this EIFA is not limited to only mouse or human MoAbs, but that MoAbs from *any* species (e.g., rat and rabbit) can be readily used.

Table 14.1. Monoclonal antibodies used in enzyme immuno-filtration.

MoAb	*Type*	*Antigen*	*Class*
T101	Mouse	Pan T-cell	IgG2a
T305	Mouse	T-cell subset	IgG2a
BA4	Mouse	HLA-DR	IgG2a
L22	Mouse	Transferrin Receptor	IgG2a
Leu2a	Mouse	Suppressor T-cell	IgG2a
Leu3a	Mouse	Helper T-cell	IgG
Leu4	Mouse	Pan T-cell	IgG
J5	Mouse	CALLA	IgG
8A1	Human	Unknown	IgM
MHG7	Human	TAA (prostate)	IgM
CLNH11	Human	TAA (cervix)	IgG
WLNA6	Human	TAA (lung)	IgM
CLNH5	Human	TAA (cervix)	IgM

General Applications of EIFA

Immuno-filtration techniques can be utilized to immobilize antigens bound to the surfaces of cells, particulate antigens, soluble antigens, and MoAbs themselves. The filter serves as the solid phase support for the antigen while the liquid phase allows for the conversion of substrate to product via an enzyme-conjugated antisera for qualitative and quantitative detection of antigens. In all cases, the immuno-filtration manifold plates have been effective in immobilizing the two major types of antigens, particulate and soluble and both are well represented in this table. Examples of cell bound antigens are those located on tumor cells and/or normal cells; particulate antigens such as detergent solubilized extracts, virus particles, and cellular debris; and soluble material usually consisting of proteins or other such ligands. With such a broad list, investigators have satisfactorily demonstrated the utility of immuno-filtration for the detection of antigens from prokaryotic and eukaryotic cells as well as cellular debris, viral particles, and macromolecules with appropriate antisera or MoAbs. Since antigens as small as EGF to the complexity of viral proteins expressed on the surface of intact cells have been immobilized by EIFA, virtually any immunogenic macromolecule may be amendable to this procedure.

In addition to the types of antigens described above, the Ig molecule itself can also be immobilized by EIFA. In the initial stages of hybridoma production it is important to identify which clones are secreting Ig. Therefore, one measures the ability of one antibody to recognize another antibody, a procedure readily amenable to EIFA technology. Since there are many separate antigenic determinants located on proteins that MoAbs recognize it may be important to distinguish one from another. The Ig molecule, in addition to the determinants found on different classes (i.e., IgG vs IgM), also expresses antigenic regions that are like fingerprints. That is, *each* Ig molecule expresses an antigenic determinant which is unique to that molecule *only* and is referred to as an *idiotype*. Within a mixture of Ig molecules there are many individual idiotypes. MoAbs have been produced against these specific Ig idiotypes and even though these reactions are extremely precise, they are still amenable to EIFA technology.

Results

In any immunoassay, quality control testing of the primary, secondary and any subsequent antisera is critical to the quantification and predictability of the desired results. Haphazard use of antisera obtained from different sources on a day to day basis, will lead to results desired only by the *Journal of Irreproducible Results*. Further, many antisera have limited shelf life when stored frozen and/or refrigerated. Many antisera manufacturers ship their reagents chilled in a solution of PBS and either glycerol or polyethylene glycol minimizing degradation of the molecular structure of the antisera as well as reducing the formation of immune complexes in solution. Titrations of each purchased antisera are completed in this laboratory upon receipt of the antisera and on a regular trimonthly basis thereafter.

Reagent Titrations

To determine the Ig concentration of an unknown hybridoma supernatant in an EIFA, a family of standard curves is generated for various dilutions of a goat antihuman antiglobulin. Serial dilutions of a chromatographically purified human IgG "captured" by several dilutions of a goat anti-human polyvalent Ig antisera. Too high a concentration of capture antisera, while increasing sensitivity for human IgG at concentrations below 200 ng/ml, is not desired because of its limited range of quantification (a function of the limitations on the optical density reading range capability of micro-ELISA readers). Concentrations of capture antisera too dilute do not provide any discriminatory

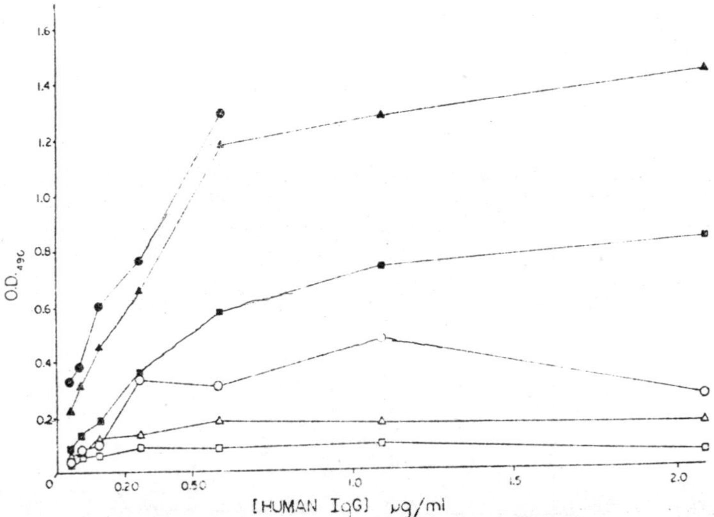

Fig 14.3. A titration of human IgG against a capture antibody affinity purified goat anti-human IgG, diluted 1/10 (●), 1/20 (▲), 1/40 (■), 1/80 (○), 1/160 (Δ), blank (□) in phosphate-buffered saline.

capability of optical density corresponding to IgG concentration. Therefore, we chose a 1/40 dilution of goat anti-human Ig (250 ng of Ig/well) because the linear portion of this curve covers the greatest range in concentration and optical density, allowing the greatest resolution of results. Binding efficiency of protein to these filters has been shown to decrease with increasing concentrations of protein. Therefore, lower concentrations allow more efficient binding of capture antisera to the filter while retaining excellent sensitivity.

Serial dilutions of chromatographically purified human IgA, IgG and IgM were bound to a 1/40 dilution of goat anti-human polyvalent Ig. This assay allows the detection of as little at 30 ng/ml of human IgA and IgG, and 100 ng/ml of human IgM. Our present human hybridoma cultures typically produce from 0.5 to 10 ug/ml of human MoAbs, significantly above the lower limits of detection. Overall, cultures are considered positive for human Ig when the optical density is 2-fold or at least 0.050 units greater than background.

The final process in our EIA involves the enzyme conjugated antibody. Like the other antibody reagents, the HRP-conjugate must also be titrated to optimize resolution. We have consistently utilized a 1/3000 dilution of the HRP-conjugated reagent.

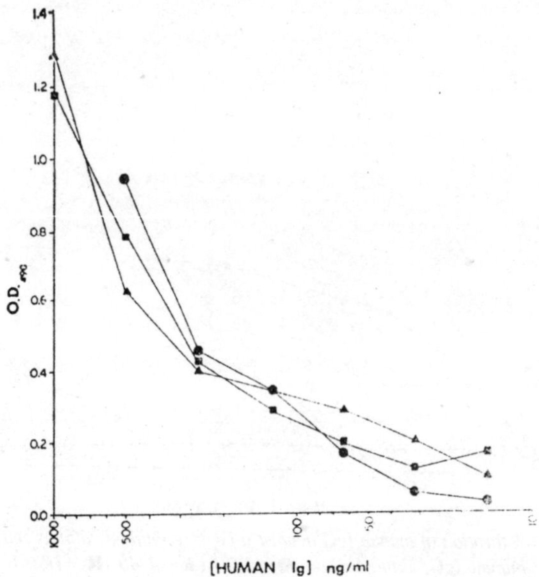

Fig 14.4. Sensitivity spectrum of an enzyme immuno-filtration analysis of human Ig. Serial dilutions of chromatographically purified IgA (●), IgG (▲), and IgM (■) were used with 1/40 dilution of "capture" antibody to establish sensitivity and standard curves.

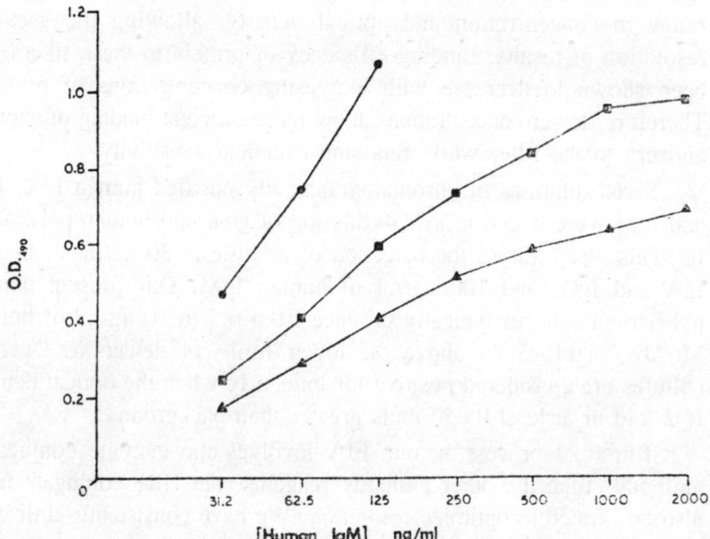

Fig 14.5. Titration of horseradish peroxidase conjugated affinity purified goat IgG anti-human IgM.

Filtration Plates Versus Polyvinyl Chloride Plates

We have compared the immuno-filtration manifold plate using the same reagents with polyvinyl chloride (PVC) plates since many immunoassays are performed with these plates. Comparisons were done with both mouse and human MoAbs using the two types of plates. Here, significantly positive O.D. readings (those greater than 0.150 U) can be obtained with 8-fold less T101 bound to cells using the immuno-filtration plates as opposed to the PVC plates. Moreover, as little as 12 ng/ml of T101 were detected with the filter plates compared to 100 ng/ml of T101 with the PVC plates.

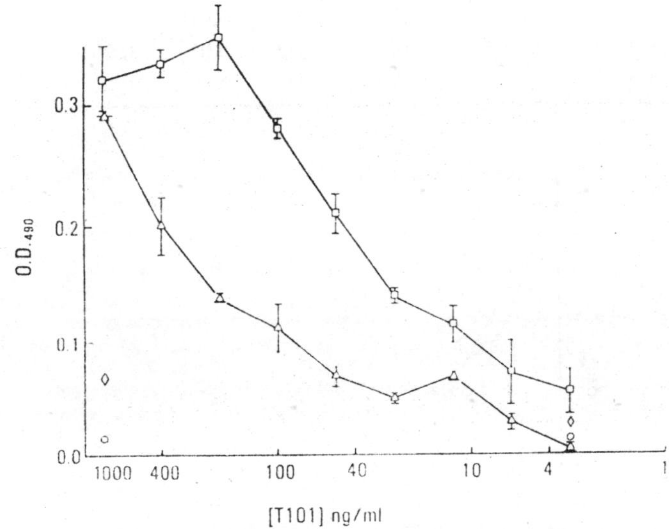

Fig 14.6. EIA comparison of immuno-filtration manifold plates (❐) with polyvinyl chloride plates (Δ). Each well of the respective plates contained 2 × 10^5 8402 T lymphocytes and was reacted with the specified concentration of T101, a murine MoAb specific for human T cells. A 1:3000 dilution of an HRP-conjugated goat anti-mouse IgG was used to develop the assay.

The human MoAb CLNH5 was also used to compare the two plate systems. It can be seen that there is essentially no difference in reactivity of CLNH5 with the cell line T293 whether tested in PVC plates or immuno-filtration plates. While background values using human MoAbs in the immuno-filtration plates are somewhat higher, this is greatly outweighed by the speed and simplicity of the immuno-filtration assay. Assay time, start to finish, was 3 hours with immuno-filtration and 6 hours with PVC plates.

Table 14.2. Titration of T101 binding to filter bound T (8402) and B (8392) cells.

	O.D.$_{490}$ Readings		
T101 (ng/ml)	*8402 T cells*	*8392 B cells*	*Filter only*
3200	0.698 ± 0.049	0.057 ± 0.024	0.021 ± 0.021
800	0.631 ± 0.051	0.043 ± 0.020	0.027 ± 0.004
200	0.616 ± 0.045	0.030 ± 0.008	0.024 ± 0.003
100	0.534 ± 0.030	0.057 ± 0.037	0.013 ± 0.013
50	0.379 ± 0.008	0.012 ± 0.004	0.018 ± 0.012
25	0.306 ± 0.001	0.035 ± 0.001	0.018 ± 0.005
12.5	0.184 ± 0.001	0.030 ± 0.018	0.053 ± 0.012
6.2	0.123 ± 0.004	0.042 ± 0.010	0.053 ± 0.035
3.2	0.076 ± 0.002	0.052 ± 0.008	0.074 ± 0.015

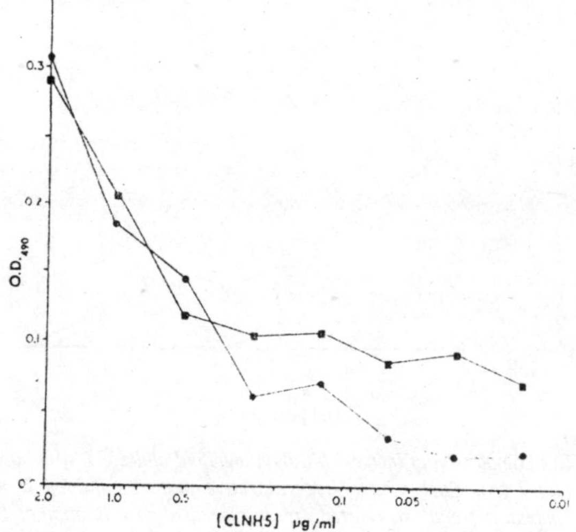

Fig 14.7. EIA comparison of immuno-filtration manifold plates (■) with polyvinyl chloride plates (●).

MoAb Immunoreactivity Against Cell Lines

Murine MoAb T101, which recognizes an antigen with a molecular weight of 65,000, reacts with all of the T cell lines. The Leu2a and Leu3a MoAbs recognize helper and suppressor T lymphocytes respectively while the J5 MoAb recognizes the cALLA antigen. It is readily evident that these MoAbs, in combination with the simplicity of immuno-filtration, allow for the careful analysis of any cell surface antigen recognized by a MoAb.

Table 14.3. Reactivity of murine MoAbs with human T cells.

	Murine MoAb			
T cell	*T101*	*J5*	*Leu2a*	*Leu3a*
8402	0.596	0.049	0.050	0.039
CEM	0.379	0.364	0.046	0.388
Molt-4	0.512	0.041	0.212	0.193
HPB-ALL	0.309	0.429	0.713	0.668

Using 2 × 10^5 target cells/well, this EIFA can also detect differences in surface density of the antigens recognized by the MoAbs. At equivalent T101 concentrations, the reactivity of T101 with 8402 cells results in higher $O.D._{490}$ readings (0.65U) than either MOLT-4 (0.54U) or CEM (0.38U), all of which are T cell lines. Similar results can be seen with the anti-T cell MoAbs Leu 4 and T305, in that Leu 4 binding to Molt-4 cells is significantly higher than to either 8402 or CEM and T305 binding is greatest with 8402 cells. The BA4 MoAb only recognizes the HLA-DR components of the B cells demonstrating an all or none reactivity pattern in that B cells express the antigen while T cells do not. The transferrin receptor recognized by the L22 MoAb is an "activation" antigen that is expressed on log-phase cells and most cell lines. All five cell lines tested were highly immunoreactive with L22.

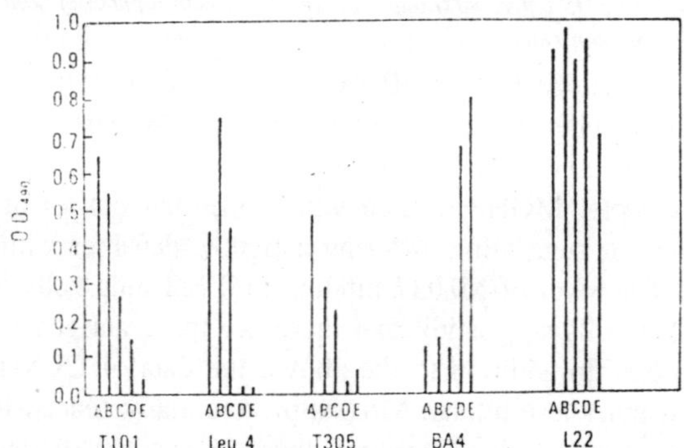

Fig 14.8. Enzyme immuno-filtration analysis of the reactivity of five murine MoAbs with a panel of human cell lines.

The EIFA immunoreactivity of human MoAbs against a panel of human cell lines is presented. Here, two human IgM MoAbs termed MHG7 and CLNH5 were reacted against a large panel of human cell types. This EIFA is well suited for screening MoAbs against a cell

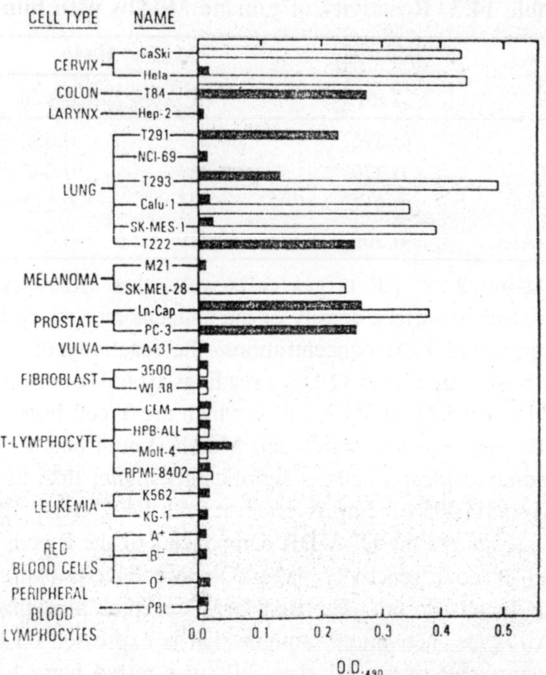

Fig 14.9. Enzyme immunoassay analysis of the human MoAbs MHG7 (closed bars) and CLNH5 (open bars) and their reactivity with a panel of normal and malignant human cells.

panel to analyze for specificity. The data demonstrates the utility of this EIFA when used in conjunction with MoAbs. One can see that the MoAbs used were able to discriminate some cell types from others. For example, MGH7 reacted with some, but not all of the lung cell lines tested, suggesting differing antigenic densities with different cell types. The level of MHG7 binding to T222 lung cells is greater than the T293 cells suggesting that there are more antigenic sites on T222 than T293. In addition to the above, the data of CLNH5 and MHG7 suggest that these human MoAbs preferentially react with tumor cells and not normal cells. Therefore, these human MoAbs may recognize a tumor associated antigen.

A comparison of the specificities of the human MoAbs MHG7 and CLNH5 shows some overlap in their reactivity with a cell panel. Since there are some cell lines that express both of the antigens recognized by CLNH5 and MHG7 questions pertaining to the nature of these antigens, their metabolic role, and whether they are of the same type can be easily addressed by EIFA. Initial studies were undertaken on the nature of the antigens recognized by MHG7 and CLNH5 by using the metabolic inhibitors cytochalasin B, puromycin, and tunicamycin, which evoke pleiotypic responses, to monitor what effect, if any, they have on antigen expression. Cytochalasin B is an inhibitor of microtubule formation and since it is a potent inhibitor of glucose transport it also inhibits protein glycosylation; puromycin is an inhibitor of protein synthesis and acts by inhibiting mRNA translation into protein; tunicamycin is an inhibitor of dolicolphosphate mediated N-asparagine-linked glycosylation. The cell line we chose to study was Ln-Cap, a prostate carcinoma cell line, since both the human MoAbs MHG7 and CLNH5 show reactivity with it. After an 18 hour incubation with the drugs, Ln-Cap cells were processed for analysis of surface antigen expression by EIFA. Puromycin inhibited the expression of the antigen recognized by CLNH5 by approximately 50%, starting at the high concentration of 50 ug/ml. Even higher concentrations of puromycin may show a greater inhibition of antigen expression. However, puromycin had virtually no effect on the expression of the antigen recognized by MHG7. Cytochalasin B decreased the expression of the CLNH5 antigen continuously throughout the concentration range used while having no effect on the MHG7 antigen. Tunicamycin decreased the expression of the CLNH5 antigen by approximately 40% while slightly decreasing the MHG7 antigen by 15%. Taken together,

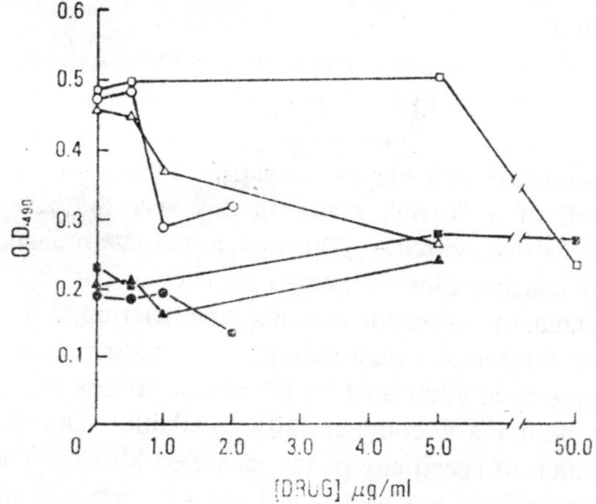

Fig 14.10. Immuno-reactivity of the human MoAbs CLNH5 (open symbols) and MHG7 (closed symbols) with human Ln-Cap prostate carcinoma cells (2 × 10^5 cells per well) after treatment with various concentration of puromycin (squares), tunicamycin (circles), and cytochalasin B (triangles).

the data suggest that these drugs decrease the surface expression of the antigen(s) recognized by CLNH5 and have no apparent effect on the antigen recognized by MHG7. This indicates that with the combined use of pleiotrophic modifiers in conjunction with antigen immobilization by EIFA, one may determine whether two or more MoAbs recognized different antigens and to analyze the nature of these macromolecules. Overall, it appears that the antigen recognized by CLNH5 may be a glycoprotein The data on MHG7 is less clear cut in that its antigen may be a protein with a long half-life or a different type of macromolecule, such as a glycolipid. Further work will be necessary to detail the molecular nature of the antigens recognized by MHG7 and CLNH5.

Discussion

One of the cornerstones of modern immunology was established upon the realization of a long sought after dream: the monoclonal antibody. Since there are many laboratories in both the public and private sectors currently using MoAb technology, it is of paramount interest to employ simple, rapid assays which can screen large numbers of hybridoma supernatants. Especially in the cases in which investigators are searching for rare antigens and consequently must screen thousands of hybridoma supernatants. To this end, we have developed an EIFA for use with MoAbs of any species in which large numbers of hybridoma supernatants need to be rapidly screened for the presence and specificity of antibody.

The immuno-filtration manifold plate is the workhorse of the EIFA. This specially designed 96-well plate serves both as a filtration manifold and as an incubation chamber. Antigen immobilization and the use of MoAbs combined with enzyme-conjugated anti-sera all occurs within a single well of a 96-well plate. In this way the assay essentially encompasses two purposes. The first is the determination of MoAb class and concentration. Chromatographically purified Ig standards provide standard curves for concentration determination. It should be noted that allogeneic crossreactivity and crossreactivity between Ig classes have been eliminated by the use of affinity purified reagents, most of which are commercially available. The second is the immuno-determination of specificity of the identified MoAbs. These immuno-filtration manifolds are well suited for the immobilization of either soluble or particulate antigenic material to screen for specificity with the test MoAbs. The use of the immuno-filtration manifolds in screening MoAbs against a cell panel shows both qualitative and quantitative differences in the amount of MoAb bound to each cell type. Also, this EIFA can be used to study subtle cell surface antigenic fluctuations through the use of compounds which evoke pleiotypic responses. Our data suggests that this assay is capable of detecting quantitative differences in the levels of cell surface antigen, in that reactivity of some MoAbs with particular cell lines gave higher $O.D._{490}$ readings than other cell lines.

It is also important to establish that short term drying of an immobilization of antigens (more important for soluble antigens than for particulate antigens) does not significantly alter antigenic structure or density of antigenic epitopes. This has been demonstrated by the data on the T cell MoAbs, in particular T101, and their reactivity to their antigens. Previously published results obtained by cytofluorographic analysis of T101 binding to several T cell types, such as T cell lines, chronic lymphocytic leukemia cells (CLL), and peripheral blood lymphocytes (PBL), indicated that T cell lines chronic lymphocytic leukemia cells (CLL), and peripheral blood lymphocytes (PBL), indicated that T cell lines had a higher cell density of surface antigen than PBL's or CLL cells. The EIFA analysis of T101 binding to T cell lines shows the same quantitative evidence of cell density fluctuations. Since these two independent assays yield similar results it would suggest that immuno-filtration does not alter the antigen.

The assay described here, when read visually, takes no more than 2-3 hours to complete. The EIFA has several distinct advantages over other assay systems, particularly RIA. These filter plates are more sensitive than equivalent PVC plate assays and don't require anywhere near the same time to complete. There is no requirement for overnight pretreatment of the assay plate in order to reduce non-specific adsorption of antibody. Washes can be completed in less than one minute without centrifugation, poly-L-lysine, or glutaraldehyde fixation. Also, cells are not lost or reduced as in the centrifugation-wash steps of other assays.

A wide variety of antigen-antibody reactions have been studied with immuno-filtration manifold plates, not only from the standpoint of the antigen, but also the antibody. In our own laboratory we have used MoAbs from several different preparations in the analysis of these reactions, such as MoAbs obtained from hybridoma supernatants, MoAbs from serum-free culture conditions, ascites, and MoAbs purified and concentrated through salt cuts ($(NH_4)_2SO_4$) and affinity chromatography. These methods of obtaining the MoAbs have essentially resulted in no difference in the ability of the antibody to recognize its respective antigen. This gives the investigator several options in how he wants to prepare the test MoAb for use. Irregardless of how the test MoAb was prepared, the power and utility of the EIFA is most evident in the screening for specificity. The test MoAbs are first screened by EIFA against the immunizing antigen, either soluble or particulate. The second and subsequent screens should be focused on the further specificity of the MoAbs to help define the reactivity profiles of the antibodies. Depending upon the interest of the investigator the EIFA can then be used to further delineate the properties of both the MoAb and its antigen.

It is very important the investigator realizes that the EIFA is not limited to the methods and applications. Indeed, the antigenic though diverse, should be thought of only in terms of examples and not limits. Our research is currently focused on the production and characterization of MoAbs to antigens associated with human cancers: more specifically, cell surface antigens. The EIFA we describe serves a useful purpose in the further understanding of the cancer problem. No matter what field of study the investigator is pursuing, if he has an antibody that reacts with a known antigen, then his research problem may be applicable to EIFA analysis. The Botanist, Agriculturist, Microbiologist, Cell Biologist et al, once made aware of the advantages of MoAb technology in conjunction with immuno-filtration, may readily find applications of EIFA in their research.

Other approaches and problems capable of being addressed by EIFA include, but are not limited to the following: analysis of MoAb affinity for its antigen; disadvantages of immunoassays on solid supports instead of in the fluid phase and ways to overcome them; the use of protein A as an immunologic reagent; immunoassays in which a mixture of MoAbs are used; and the use of MoAbs to study epitope specificity and biotin-avidin applications. Furthermore, there are a wide range of applications in using immuno-filtration in DNA hybridization experiments in which radiolabled probes can be used to detect fragments of the DNA under study.

Conclusions

The EIFA we describe in this report utilizes a specially designed 96-well microtiter plate that serves as both an incubation chamber and as a filtration manifold. This EIFA provides a rapid, simple, and sensitive method of detecting MoAb class, concentration, and specificity. In this assay either soluble, particulate, or whole cell antigens are immobilized onto filters with subsequent analysis by EIFA sing appropriate MoAbs. This EIFA requires small volumes of antisera, few target antigens, and can be completed in as little as two hours. In addition, this assay is well suited for the rapid screening of large numbers of hybridoma supernatants and is more versatile and faster than standard PVC plates.

15

Application in Medicine

Better than many other areas of "biotechnology," the short history of interferon research, since their discovery in 1956, illuminates the problems and promise of the new technology as applied to biomedical practice.

Comprehensive reviews of the biology of interferons have appeared at the rate of several per year. To these, the reader is referred for supporting detail and many further references. Since the late 1960's, the field has grown so explosively that a truly comprehensive review by a single author no longer seems possible. What follows, therefore, is an attempt to overview important features of the biology of interferons which relate to their successful manufacture and application in medicine.

The areas where significant problems remain are: (1) large-scale production of interferons; (2) purification and characterization of these, yielding lyophilized materials with useful shelf-life; (3) comparative pharmacodynamic characterization of the various interferons *apart* from pharmacologic effects in specific diseases; and (4) comparative pharmacologic studies of the many interferons with reference to particular diseases. To these problems should be added our virtual ignorance of (5) the occurrence of *endogenous* interferons in various disease states, (6) the long-term effects of therapy with various interferons and (7) the frequency of situations in which interferon itself mediates pathogenic mechanisms.

It is evident that the first two problems are technical problems relating to manufacturing methods, while the remaining problems are classical *pharmacologic* problems attending the introduction of *any* new drug. As such, these problems are *not addressed* by recombinant DNA or hybridoma technology, nor by other recent developments in commercial "biotechnology."

Nomenclature of Interferons

Antigenic Classes of Interferons

"Interferon" was discovered as an antiviral activity, without reference to antigenic type or tissue of origin. It was later observed that interferons produced from lymphoid, as opposed to fibroblastoid or epithelioid cells, are not cross-reactive antigenically. They are also distinct regarding heterospecific activity using bovine or porcine cell cultures and in certain gross chemical properties. Therefore, the distinct interferons were named according to tissue or origin—"fibroblast" or "leukocyte." Subsequently, it was found that lymphoid cells are capable of synthesizing "fibroblast" interferon, and that fibroblastoid cells are capable of synthesizing "leukocyte" interferon, in response to viral infection. Thus, the terms "fibroblast" and "leukocyte," as applied to human interferons, denote *antigenic classes* of interferons and *not* the cell type of origin. For this reason, the human interferon classes were redesignated alpha and beta (leukocyte and fibroblast, respectively). In the mouse interferon family, the terms "alpha" and "beta" refer, respectively, to the acid-stable interferons of higher and lower electrophoretic mobility in SDS-PAGE gels, and the terms "fibroblast" and "leukocyte" have no meaning (virtually all mouse cells, regardless of tissue of origin, produce both species of interferons). In both human and mouse systems, the most effective inducers of these interferons are viruses and double-helical RNA's of natural or synthetic origin. Together, acid-stable alpha and beta interferons (human and mouse) are the "classical" interferons. *By definition*, alpha and beta have been regarded as acid-stable interferons, though acid-labile *alpha* interferons (e.g., human and bovine) have been described.

"Lymphoblastoid" interferons are those derived from continuous lymphoblastoid (generally B-cell) lines, which are handled much as for leukocytes as regards interferon induction and purification, but which obviate the requirement for fresh blood or leukopheresis. Various lymphoblastoid cells produced beta as well as alpha interferons, or combinations of these.

As early as 1965, a species-specific antiviral activity, with some of the properties of interferon, was discovered in the supernatant of leukocyte cultures which had been simulated with lectins or other nonspecific mitogens. The antiviral activity, however, is acid-labile. These acid-labile interferons have been denoted "immune." "T-" and "Type II" interferons. They were subsequently shown to be antigenically distinct from either alpha or beta interferon in both human and mouse systems, and now have been designated "gamma" interferons.

How Many Interferons are There?

Partial amino acid sequences have been published which establish the distinctness of human and mouse alpha and beta interferons. A combination of studies employing screening of genomic DNA fragments, cytogenetic studies of the segregation of interferon loci, and biochemical studies of the proteins and their messenger RNAs, presently suggest (in the human system) that there are approximately fifteen IF-alpha, at least three IF-beta, and at least one IF-gamma locus (exclusive of allelic variation within these). For 4-5 of the alpha, two of the beta and the single gamma sequence, some information is available from study of the activities of the gene products themselves. Indirect evidence suggests that there may be more than three beta sequences. If less closely related sequences are discovered, for example, using less stringent hybridization methods, further effort will be required to study the corresponding gene products and to determine what significance, if any, these have under physiological conditions.

Properties of Interferons

Physicochemical Properties

Interferons are extremely potent natural substances comprising a family of proteins with molecular weights in the range 20-4 $\times$ 10^3 d. In some cases, interferons are glycoproteins and probably contain other post-translational modifications including phosphorylation, disulfide linkage to external substances, hydrophobically-bound moieties and, possibly, quaternary structure. The chemical characterization of interferons has been impeded, in the past, because of (1) the exceedingly small amounts of interferon protein available (2) their contamination by relatively huge amounts of non-interferon proteins, hindering purification, (3) the multiplicity of distinct interferon gene products and possible post-synthetic modifications (i.e., micro-heterogeneity) of these, and (4) their chemical properties, such as adsorption to other substances or instability of the sulfhydryl present in human IF-beta. These restrictions are such that classical methods of purification and amino acid sequencing did not yield partial sequence data until 1979, and have progressed slowly since.

Partial amino acid sequence data, however, allowed the synthesis of corresponding oligonucleotide sequences and the use of these as hybridization probes in screening recombinant plasmids for the present of interferon sequences. Such plasmids have been generated by cloning restriction fragments generated from genomic DNA, as well as from complementary DNA synthesized enzymatically from cellular RNA containing interferon mRNA sequences. The genomic sequences for human alpha and beta interferons are themselves expressed as functional interferons in bacteria, since introns are absent.

The techniques of polynucleotide sequence determination and manipulation have provided a great impetus to studies of the chemical nature of interferons. Interferon structural sequences, particularly human alpha, beta and gamma interferons, have been transferred to, and expressed in, bacteria, yeast and animal cell cultures. The manipulation of recombinant sequences has made possible (1) greatly increased expression of interferon sequences in recombinant organisms, (2) hybrid interferons containing amino acid sequences derived from different parent interferons and (3) interferons containing amino acid sequences intentionally modified for greater chemical stability. Sequencing of these genes has provided the amino acid sequences of the family of human and mouse interferons and has allowed comparison of these at the primary structure level. While not yielding data directly concerning secondary and tertiary structure, such studies have allowed identification of highly conserved sequences which are probably involved in maintenance of functional tertiary structure or in possible active sites.

As yet, little is known of structure-function relationships for any interferon, nor of the tertiary and possible quaternary structure of any interferon. Indeed, the conditions which have generated a consensus concerning molecular weights of interferons (denaturing conditions) eliminate native tertiary and quaternary structure. The recent preparation of a human IF-alpha in crystalline form should facilitate crystallographic study of higher-order structure.

Criteria for Interferons

Interferons were discovered, and until 1979 were described purely, as biological activities. Classically interferons were described according to a set of empirical criteria. By 1980, however, so many exceptions to these criteria had been discovered that the consensus was revised: interferons were defined as "... proteins which exert nonspecific antiviral activity at least in homologous cells through cellular metabolic processes involving synthesis of both RNA and protein".

For a number of reasons, this definition is grossly unsatisfactory. First, the "nonspecific" antiviral activities of interferons vary greatly from one *virus* to another, and are known to depend also upon the host cell or animal employed, the method for quantifying "antiviral" activity, and in some cases the genotype of the animal or cell strain employed. Second, the antiviral activity is properly regarded as one manifestation of a set of alterations of cellular metabolism induced by interferons; conversely, other effector proteins may be discovered which have antiviral effects mediated by mechanisms distinct, at least in part, from those described for interferons, and which may not be regarded ultimately as interferons. Third, the *specific* antiviral activities of the distinct human interferons vary widely (3-4 orders of magnitude) within human (i.e., homologous) cell strains, and for some of these the activity is far *greater* in heterospecific cells. Fourth, although classical studies of the interferon-induced antiviral state clearly demonstrate the requirements for RNA and protein synthesis, the same cannot yet be said for other metabolic effects of interferons, and the relationships between these and the antiviral effect, per se, are obscure. In short, the definition of interferons based upon empirical biological properties has become less rigorous as more becomes known, and will certainly be subject to revision in the future.

However, beginning with the publication of partial amino acid sequences in 1979, interferons could be defined also according to primary structure (amino acid or nucleotide sequences). Nucleic acid sequences, properly speaking, define *structural* sequences of the corresponding proteins which, in the case of the human interferon family, are mostly *inferred* to exist and which have not been characterized directly. Thus, although the recombinant methods have yielded unambiguous description of the interferon gene products to which they have been applied, they have greatly compounded the problem of defining interferons operationally. Recombinant, cytogenetic and biochemical studies, as mentioned, have expanded the catalogue of interferons from three to perhaps twenty in the human system; for many of the individual gene products, little is known concerning their biological properties, effects on the actions of other similar molecules, nor of their mechanisms of induction and regulation of expression. The multiplicity of interferon sequences in the human genome, inclusive of pseudogenes, raises new questions concerning the identity of interferons. It is not clear how far abreast of the classical interferons such studies may take us. It is possible that there exists a much larger family of interferon-related sequences, and many corresponding gene products, of which we know nothing at this time. Formally, the demonstration of perhaps fifteen distinct IF-alpha loci, using hybridization, does not prove that each of the gene products is an interferon by the usual biological criteria.

Thus, the "definition" of an interferon cannot be rigorous at this time, and remains chiefly a pedagogic aim. It is likely that our views will be strengthened by rigorous chemical and biological characterization of the many interferons, and equally likely that more proteins, beyond those currently recognized, will properly be regarded as members of this family of substances.

Biological Properties

The biological actions of interferons arise from their interaction with the surface of cells. At this time it cannot be ruled out that interferons have biphasic mechanisms of action, like insulin, which acts both upon surface receptors and is also transported into the cell. At least some of the antiviral actions appear not to require that the molecule be transported into the cell and other evidence suggests that the antiviral action must be triggered from the exterior of the cell. Not merely in certain mechanisms of action, but in many apparent chemical properties, interferons resemble other members of a group of polypeptide hormones which includes insulin, follicle-stimulating hormone (FSH), luteinizing hormone (LH), thyrotropin (TSH), somatomedin, somatostatin and others, as well as a group of bacterial (diphtheria, cholera) and other toxins (ricin, abrin).

The so-called "nonantiviral" or "anticellular" actions of interferon have been described virtually since interferons were discovered. The terminology is cumbersome because *all* known actions of interferons are exerted by their actions upon cellular metabolism. These include:

1. "Priming" of the *synthesis* of interferons by interferon pretreatment, resulting in earlier onset of synthesis after induction, increased rates of accumulation and higher amounts of interferon produced.
2. Growth inhibition in many types of cells *in vitro* and *in vivo*.
3. Generalized inhibition of cellular motility and gross alternations in cellular adhesion, volume and content of DNA and protein *in vitro*.
4. Inhibition of *maturation* of terminally-differentiated cell types including granulocytes, the monocytemacrophage line and erythropoietic cells *in vitro* and *in vivo*.

5. Inhibition of continued expression of differentiated biochemical functions, such as hepatic hemoprotein metabolism, the cytochrome P450 detoxification chain, adipose conversion, or steroid-dependent induction of gene expression.
6. Enhancement of the toxicity, and production of certain prostaglandins, caused by certain viruses and double-stranded RNA.
7. Gross alterations in intracellular levels of cyclic nucleotides and other intracellular messenger substances, including protein kinase-phosphorylase systems having regulatory functions.
8. Many types of effects on the expression of immune responses, including effects on antibody synthesis which may be stimulatory or inhibitory, as well as effects upon cell-mediated immune functions including immediate and delayed hypersensitivity, cytotoxic T-cell function (and, independently, target-cell susceptibility), natural killer cell activity, and the activity of phagocytic cells.

At present, interferons are still quantified almost exclusively by serial endpoint dilution assays of the antiviral activity using convenient combinations of cell lines and viruses which are rapidly cytolytic to these. Many other types of assays have been devised, based upon various other activities of interferons which can be quantified *in vivo* and *in vitro*. Immunochemical assays employing monoclonal anti-interferons and pure, radio-labeled interferons are now being devised and will come into widespread use in the future. Such assays will help to provide information of diagnostic and prognostic value in situations where endogenous interferons are preclinical or clinical markers for various diseases. In spite of the imminence of immunochemical assays, which are *physical* assays employing cell cultures will not diminish since biological assays are a necessary component of quality control and in pharmacodynamic monitoring of clinical studies.

Production of Interferons

Prior to the construction of recombinant microorganisms, interferons were derived exclusively from animal tissues or cell cultures. Historically, animal cell cultures have been inefficient and costly, with typical specifications for crude interferons of 10^3-10^4 IL/mL and 10^3-10^4 IU/mg, *secreted* from the cells into the medium. This represents a few nanograms of interferon protein per milliliter, requiring purifications of 10^5- to 10^6-fold to achieve homogeneity.

The synthesis of interferons is stimulated, *in vivo* and *in vitro*, by infection with viruses, or by treatment with a variety of nonviral substances, many of which are high molecular weight polyanions. In all of the mammalian and avian systems which have been studied, the principal inducers of "classical" interferons (alpha and beta interferons in the human and mouse systems) all comprise natural or synthetic double-helical ribonucleic acids, which are either added exogenously or are produced during virus replicative cycles. In cells of lymphoreticular origin and *in vivo*, in addition to viruses and dsRNAs, interferon synthesis can be induced by a group of polyanionic substances including polycarboxylates, polyacrylates, polyphosphates, oxidized polysaccharides and bacterial lipopolysaccharides. A group of low molecular weight substances are interferon inducers *in vivo*, including kanamycin, cycloheximide, tilorone, N,N´-alkyldiamino-propane derivatives and other substances.

Cell Cultures

In early studies, interferons were induced exclusively by viral inducers. Human leukocyte as well as nonhuman interferons are still produced efficiently using the paramyxovirus inducers (Newcastle disease or para-influenza type 1 viruses). Fibroblast interferon is usually prepared in human fibroblast cultures using polyinosinic acid-polycytidylic acid duplex, in combination with inhibitors of RNA and protein synthesis, in the "super-induction" method. For nonrecombinant cell cultures, these inducers remain the most efficient. Their action, moreover, can be enhanced in various production systems by additional substances including 5-bromode-oxyuridine, sodium butyrate, theophylline, polycations such as diethylaminoethyl-dextran, insulin, cyclic nucleotides and their N^2,O^6-dialkanoyl derivatives, glucose, ascorbate, p-mercuribenzoate, and agents which modify membrane fluidity or permeability, such as dimethylsulfoxide, amphotericin B and cytochalasin B.

Preparation of human IF-alpha has necessarily employed leukocytes prepared from the large quantities of fresh blood. Blood cannot be viewed realistically as a suitable source, either quantitatively or qualitatively, because of problems arising in collection, storage, contamination with infectious agents, and because of the labor-intensive nature of the procedures for induction and purification of interferons. Whereas primary diploid human fibroblast strains have been the preferred source of human IF-beta, they have a limited lifespan *in vitro* and are not likely to supply a large demand for this material. Continuous cell lines, while not currently recognized as suitable substrates for the manufacture of biologicals, do have the advantage of an indefinite lifespan *in vitro* as well as other advantages. In all cases, the production of interferons from (nonrecombinant) cell cultures requires treatment of the cultures with interferon inducers, such as poly I:C or viruses, which are costly (and, in some case, hazardous) by comparison with recombinant bacterial strains which can be induced with simple, unrelated substances or which can produce interferons constitutively.

IF-gamma is produced by peripheral or spleen lymphocytes in response to nonspecific mitogens (e.g., phytohemagglutinin, pokewood mitogen, concanavalin-A), specific mitogens (antigens to which the cells have been sensitized), or by treatment of the cells with (1) antisera against the cell surface, (2) galactose oxidase or other enzymes capable of derivatizing the cell surface or (3) a calcium ionophore. In other words, synthesis of IF-gamma appears to be triggered by any agent capable of altering the lymphocyte membrane, whether mechanically, covalently or electrochemically.

More recently, human IF-gamma has been prepared from peripheral lymphocytes, which are induced using phytohemagglutin or staphylococcal enterotoxin A, and which are incubated in medium containing phorbol esters to increase the yields of interferon. Such methodology is subject to all of the criticisims of leukocyte IF-alpha, mentioned above. In addition, phorbol esters are extremely hazardous substances (tumor promoters) and the mitogenic stimulation of lymphocytes can result in shedding of T-cell tropic lymphoma/leukemia viruses by the cells. Except as discussed below, both of the blood cell-derived interferons, and all other blood-derived products, are in every case potentially contaminated with viruses and viral nucleic acids arising from the cells.

Realization of the potential of cell culture technology depends upon several areas of further endeavor. Most of the cell types amenable to traditional cell culture methods are derived initially from neoplastic tissue or frequently acquire "neoplastic" properties *in vitro*. Nutrient media for culture of animal cells have evolved very little since the 1950's, and are not optimized for cell culture at densities exceeding those for traditional monolayer or spinner cultures (about 10^6 mL^{-1}). Further work is required to develop cell substrates which preserve differentiated functions and which can be grown in defined media free of all exogenous protein.

The growth of many useful cell strains requires attachment of the cells to a solid substratum. Even for cells which are capable of growth in suspension, manufacturing advantages can accrue from the use of immobilized cultures. The development of stationary as well as suspended cultures employing microcarriers has increased culture density by a factor of about ten during the past decade. In our view, further increases in density and efficiency of manufacture, amounting to perhaps one hundredfold, will be achieved by employing microcarriers in continuous-flow systems containing fluidized beds.

Our own work has proceeded on the assumption that cell cultures are a resource of great potential, by no means eliminated from consideration as sources of interferons, vaccines and other substances on the scale required for widespread clinical use. Our efforts have therefore proceeded, and considerable future work is required, along the following lines:

1. The cytogenetic manipulation of human and animal cell lines, so as to construct cell lines with desirable properties regarding growth, ease of induction of interferon synthesis, susceptibility to enhancers of interferon yields, identity of the interferons produced, or amenability to culture at high densities and in serum-free, defined media.
2. By employing culture methods based upon commercially-available microcarriers, production schemes have been devised for human and animal interferons with efficiencies about 5,000-fold greater than those typical of the original methods, cited above. That is, crude extracellular interferons are produced with $>10^6$ IU/mL and 10^6-10^7 IU/mg. In other words, on a *pilot scale*, cell cultures are capable of yielding individual interferon gene products with a cost-efficiency equal to that for recombinant bacterial methods.

The effective sterilization of biologicals manufactured using cell cultures has been a serious problem. Obviously, this pertains equally to products, including interferons, manufactured from blood or blood fractions. In our view, improved diagnostic methods for screening of cell cultures, or blood, will be of limited use in improving this situation. The same can be said for methods designed to remove putative contaminants. It would seem that methods enabling the *inactivation* of putative contaminant organisms, or nucleic acids, will circumvent the problems of limited sensitivity which attend any screening method, or limited efficiency in any purification method. These problems are closely related to policy which determines the suitability of cell culture substrates for the production of biologicals for parenteral use. At present, the fear of potential contamination of biologicals, by viruses or nucleic acids which are infectious or oncogenic, is the principal basis for regulatory policy. Consequently, few types of cell cultures are viewed as acceptable in manufacture, and these are restricted to diploid human fibroblasts which have been regarded as "normal" cells. The criteria for "normalcy," however, are less than satisfactory.

The most important consequence of the fears expressed above is the inability to employ continuous cell lines, particularly "genetically-engineered" strains, in manufacture. The problem is one of effective inactivation of adventitious infectious or oncogenic agents arising from the cell substrates. For this reason, we have employed photochemical inactivation of nucleic acid function, using various psoralen derivatives and longwave ultraviolet radiation, for sterilization of interferons, vaccines and other substances produced from cell cultures. Such methods are capable of completely inactivating any virus or nucleic acid arising from the cell substrates. Therefore, these methods are expected to remove the fears upon which regulatory policy is currently based, and to bring into use the great potential which cell culture technology appears to hold.

Recombinant Microorganisms

Principally because of the promise of interferons, as magnified and distorted in the press during the late 1970's, considerable excitement attended the announcement by Biogen SA, in January of 1980, of the cloning and expression of human IF-alpha sequences in *E. coli*.

Currently, mammalian interferons can be manufactured using recombinant protists with typical yields of 10^6 IU/mL and purities of 10^6-10^7 IU/mg protein. The interferons, however, are not secreted into the medium but must be extracted from the cells. As advances occur enabling better efficiency of production by recombinant organisms, it is estimated that the concentration and purity of crude recombinant interferons may increase over those cited by an order of magnitude. There remain, however, several additional considerations:

1. Recombinant strains are required which secrete the interferon into the culture medium. This will eliminate the necessity for extraction of the interferons from what is essentially a concentrated solution of lipopolysaccharides; conversely, gram-positive organisms may be preferred to minimize the extent of contamination by endotoxins.
2. Recombinant bacteria are not expected to attach carbohydrate moieties to those interferons which are normally glycoproteins. The carbohydrate is not required for antiviral activity, but its presence is expected to influence the pharmacokinetic properties of the interferon, may affect other "nonantiviral" activities of the

molecules, and may affect the details of purification and physical properties of the product.

3. Recombinant technology allows the unambiguous identification and production of individual interferon gene products in unlimited quantities; conversely, the spectrum of molecules present in *non-recombinant* interferon preparations must then be re-examined by reconstruction experiments. Moreover, as the number of "natural" interferons, "hybrid" interferons and otherwise modified interferons increases, each must be studied individually and comparatively, pharmacologically speaking. The possibilities of "second order" effects, arising from cooperative or antagonistic actions of individual species, rapidly become mind-boggling. Yet there is considerable evidence both for heterogeneity of "native" (nonrecombinant) interferons produced *in vivo*, as well as for cooperative effects among the components. Here, then, is an area of study which may be practically unlimited.

Methods of purification based upon sorption to organic ligands, such as cibachron blue dye, are being applied on a manufacturing scale. Methods based upon immunochemical affinity are expected to supplement these within a few years. Neither of these methods is without drawbacks: for the former, the slow release of the dye, which is potentially carcinogenic; for the latter, the slow release of immunoglobulin which is potentially antigenic. Thus, neither of these is a preferred final step in purification of materials for parenteral use. Using HPLC, sufficient human IF-alpha has been prepared to permit crystallization; it remains to be seen how quickly HPLC or newer derivative methods can be scaled up, and what the costs will be.

Future progress in the large-scale purification of interferons is likely to proceed along the following lines:

1. One wishes to obviate the need for extraction of intracellular interferon, and to maximize the specific activity of crude interferons accumulated in the extracellular medium. This is done by minimizing the necessity for exogenous protein in such media, by minimizing the concentration of cell-derived non-interferon proteins, and by maximizing the rate at which interferons are secreted and removed from the culture milieu.
2. One further wishes to minimize the necessity for concentration prior to further purification, or conversely, to achieve partial purification in the course of concentration, as for example, when

interferon can be removed from a continuous-flow culture reactor using sorbents which are relatively specific.

3. One seeks purification schemes with a minimal number of steps, and with compatibility between steps which obviates the necessity for intervening dialysis or other handling.
4. One further seeks purification schemes wherein the last step, at least, does not employ sorbents which can release toxic, carcinogenic or antigenic substances into the final product.
5. Finally, it is desirable to employ schemes in which the final step yields material in a vehicle which is physiologically compatible, if not pharmacologically advantageous.

To date, the preparation of highly purified interferons in commercially useful form is practically uncharted territory. The principal exception has been a group of human, mouse, rabbit and rat interferons available commercially since 1977. This experience has shown that highly purified interferons can be lyophilized in a form which is sterile, stable during transport and ambient temperatures, and has good shelf-life (indefinite at refrigerator temperature). From the manufacturing point of view, lyophilized products, which are stable at ambient temperature, are immensely preferable to materials in solution which require frozen storage and transport. Such materials have employed, as vehicle, a number of simple isotonic salt solutions as well as more concentrated electrolytes which confer desirable advantages, for both manufacturing and pharmacologic reasons, upon the interferons. Our experience suggests that this area is rich in possibilities, not only those mentioned above, but in further modifications of the suspending vehicle which allow manipulation of the activity pharmacodynamically. It is clear that such methods, already developed for manufacture of research materials, will prove useful in the pharmacologic application of interferons in the future.

It should be mentioned that, at present, commercial products derived from recombinant microorganisms (or from cell cultures, for that matter) are in the earliest stages of licensing. At this stage, major problems seem unlikely, but unforeseen minor setbacks and diversions are certain to occur.

In summary, there is no doubt that recombinant microorganisms have become useful sources in the manufacture of interferons and related substances. The problems attending the production of large quantities of a given interferon molecule, as well as novel molecules, have largely been solved by recombinant techniques. This will enable the eventual characterization of these molecules by classical studies of structure and function; these, no doubt, will proceed rapidly during the 1980's. It seems equally likely, however, that cell culture-based technology will remain important (1) as a source of "native" interferons for comparative studies with recombinant interferons, and (2) as an alternate source of "microorganisms" and culture methods for the application of recombinant methods.

PURIFICATION OF INTERFERONS

Until the late 1970's, all methods for the concentration and purification of crude interferons (produced from cell cultures) employed traditional techniques derived from work with other proteins. Many of these remain extremely useful. For example, concentration of interferons (without significant purification) can be achieved effectively by ultrafiltration, ammonium sulfate precipitation, or adsorption to zinc hydroxide precipitate or other inorganic sorbents formed *in situ*. Purifications, typically from ten- to one hundredfold per step, are achieved by sorption chromatography on a variety of inorganic and organic ligands (e.g., controlled-pore glass beads, immobilized cibachron blue or similar chromophores, polynucleotides or hydrophobic ligands) or by gel filtration chromatography. Polyacrylamide gel electrophoresis is capable of 1000-fold or higher purifications, but is likely to remain useful chiefly as an analytical tool, or as a preparative tool limited in scale to research use.

Purification of interferons by immunoaffinity chromatography has been effective for several types of interferons, with purifications of 1000- and higher in a single step. Immunoaffinity chromatography employing high-affinity monoclonal immunoglobulins is expected to provide an efficient method for purification and resolution of the various interferon species in the future.

The imminent accessibility of pure, crystalline interferons, produced from recombinant organisms, will have magical consequences for further research. As mentioned, crystallographic analysis will provide essentially the first information of any sort concerning the tertiary and quaternary structure of interferons. This will advance our knowledge of the structure-function relationships of interferons, and this, in turn, will clarify mechanisms of action. The accessibility of pure, radio-labeled interferons will add a new dimension to the identification and quantification of these substances, and will also facilitate ultrastructural studies of the cytological mechanisms involved.

PHARMACOLOGIC CHARACTERIZATION OF INTERFERONS

Pharmacodynamics

In comparison with well-established chemotherapeutic substances, knowledge of the pharmacodynamics of interferons is relatively crude. There are few detailed studies of pharmacodynamics in humans or animals. Such studies, in humans, have been limited mainly to leukocyte interferon (unfortunately, to a much lesser extent for fibroblast interferon) since leukocyte interferon, prepared using Cantell's methods, has been used in most clinical trials. At present, recombinant alpha and beta interferons are in initial clinical testing; therefore, it is likely that a period of some years is required to accumulate detailed knowledge of the requisite pharmacodynamics for several types of interferons.

The drawbacks of conventional leukocyte-derived interferon have been apparent since the outset; it is largely the dedication of Kari Cantell and Hans Strander which was responsible for bringing leukocyte interferon into clinical research. The procedure requires a continuous large supply of fresh peripheral leukocytes obtained, preferably, by leukopheresis. The interferon is stimulated with paramyxovirus inducers, concentrated and partially purified in a series of steps, yielding material at 1-10 $\times$ 10^6 IU/mL and 10^6-10^7 IU/mg. The chronic administration of such material at doses of about 10^7 IU/week has been well tolerated with generally minor side effects (low-grade fever, malaise, lethargy, thinning of hair, mild immunosuppression and occasional aplastic anemia).

Since the initial use of human leukocyte interferon in humans, the *maximum* doses have been 1-10 $\times$ 10^6 IU per dose, administered intramuscularly, at frequencies up to b.i.d. (i.e., up to 20 $\times$ 10^6 IU/day). If a broad conclusion can be drawn, it is that doses in the range employed to date lie at the low end of the therapeutic range, and are also the doses at which side effects begin to be apparent. In other words, the therapeutic index for leukocyte-derived interferon, used in most clinical trials to date, is uncomfortably low. This toxicity derives from viral components and, perhaps, substances arising from the leukocytes. Recently, cell culture-derived human IF-beta, as well as recombinant IF-alpha and IF-beta, have been tested in dosages of $>10^8$ IU/dose with good acceptability in the short term, and without the side-effects mentioned above. The administration of an equivalent amount of the conventional leukocytederived interferon would have catastrophic consequences.

Initial comparative studies of human leukocyte and fibroblast interferons showed that fibroblast interferon is not well absorbed from an intramuscular site, whereas leukocyte interferon is well-absorbed. To some extent this observation has served as the rationale for the emphasis in clinical trials on leukocyte interferon. The rationale is unfounded since (1) fibroblast interferon can be modified so as to be absorbed from an IM site, (2) fibroblast interferon can be administered by other routes, and (3) fibroblast and leukocyte interferons, numbering among them perhaps 20 different gene products, are expected *a priori* to have different pharmacodynamic properties; conversely, such differences, when understood, may be exploited in the differential indication of the various interferons.

The demonstration of approximately twenty distinct interferon structural sequences within the human genome (and possible allelism within any of these) is stunning in its implications for the use of interferons therapeutically. An extension of the foregoing argument implies that one is dealing with approximately twenty distinct pharmacologic entities; to these must be added *mixtures*, as are often produced *in vivo*, of the individual gene products, since these appear to act cooperatively among themselves and in concert with other substances such as lymphokines and the peptide hormones or toxins mentioned above. In some instances, synergistic effects are considerable: for both the human and mouse systems, IF-gamma and IF-alpha *or* IF-beta show a synergy amounting to 30- to 100-fold as regards antiviral activity. Potentially, such interactions may modulate other "nonantiviral" activities of interferons as well.

Thus, even before admitting allelic sequences and hybrid or otherwise modified recombinant interferons, and before considering any "second order" pharmacologic effects (synergistic or antagonistic effects), we cannot speak of "interferon" as a substance, but rather approximately twenty substances, each of which must eventually be characterized, individually and comparatively.

Recombinant Interferons

Initial clinical studies with recombinant human IF-alpha suggest that this material is more antigenic than comparable amounts of IF-alpha produced from leukocytes. Is this because the highly purified material is intrinsically more antigenic? Or is increased antigenicity characteristic of this particular gene product? Or is the increased antigenicity related to contamination by trace amounts of lipopolysaccharide (an effective adjuvant) arising from the bacterial source? Or is the increased antigenicity reflective of the *deletion* of other gene products, which may act antagonistically or immunosuppressively, and which are present in the "native" material? The answers are not yet available; conversely, it must be stressed that the solutions to the problems of production, on one hand, have created a new set of problems on the other hand, i.e., the necessity for establishing detailed identity or differences between "native" and recombinant interferons.

Unfortunately, it is still too early to answer a fascinating question: as evolutionary relationships between interferon genes and other descendants of common ancestral sequences are uncovered, what are the corresponding gene products and their roles in regulation, if any, similar to those of interferons? The extent of homology between the human alpha and beta sequences suggests that these genes diverged several hundred million years ago, roughly coincident with the emergence of vertebrate phyla. What were the ancestral invertebrate proteins and their functions? The existence of 15-20 alpha and beta genes and pseudogenes shows that, since that time, interferon sequences have continued to duplicate and diversify at a remarkable rate. Are all of these sequences (excluding pseudogenes) actually expressed *in vivo*? And under what conditions, i.e., differentially? Are there indeed fifteen distinct *pharmacologic* entities? Why so many—what adaptive advantages, if any, does this confer? Most interestingly: in time we shall learn the mechanisms by which the sequence redundancy and diversity are generated, and whether interferons may illuminate processes in the evolution of other families of sequences. The lack of introns in the genomic sequences suggests that the mechanism of duplication could involve reverse transcription from messenger RNA's, but is also compatible with recombination of genomic sequences which already lack introns.

Immunotherapy

All interferons appear to possess a wide variety of actions upon the cells, and their functions, which comprise the immune and inflammatory responses.

The effects of interferons on antibody synthesis may be stimulatory or inhibitory, depending on the nature of the antigen, the timing and dose of interferon, and whether the response is primary, secondary etc. However, for certain antigens only a suppressive response is found. In this suppression IF-gamma appears more potent, per unit antiviral activity, than IF-gamma appears more potent, per unit antiviral activity, than IF-alpha or If-beta. The immunoregulatory effects of interferons on antibody synthesis could be mediated by any of the target cell types involved: memory B cells, antibody-synthesizing B-cells, helper T- and suppressor T-cells, and possibly NK cells and macrophage.

Among the immunologic actions of interferons which have been reported are:

1. Alteration of cell surface properties, as evidenced by promotion of patching and capping of membrane antigens and redistribution of cytoskeletal components. Such observations probably point to the generalized *cytokinetic* effects of interferon, but are paramount in a discussion of immunoregulatory effects since it is by intercellular recognition and/or contacts that all immune responses, afferent and efferent, occur. Interferons enhance cell-surface expression of major histocompatibility antigens, with implications for many cell-mediated mechanisms involving recognition of these.
2. Interferons are reported to modulate differentially the expression of immunoglobulin receptors on the lymphocyte surface.
3. Interferons inhibit lymphocyte proliferation after activation by lectins, antigens and allogeneic cells.
4. Interferons enhance the cytotoxic phase in mixed lymphocyte cultures by enhancing target-specific T-cell cytotoxicity as well as spontaneous (SK) and natural (NK) killer-cell activities. Interferons also enhance antibody-dependent cytotoxic lymphocyte activity.
5. Interferons enhance the phagocytic activity of macrophages and polymorphonuclear cells, causing increases in spreading and enzymatic activities associated with activation; as mentioned above, interferons inhibit monocyte and granulocyte maturation and bone marrow poiesis.
6. Interferons enhance histamine release by basophils in response to specific antigens or antibody reacting with the basophil membrane.
7. Interferons inhibit both the establishment and maintenance of delayed-type hypersensitivity.
8. Interferons inhibit graft-vs.-host disease, allograft rejection and cytomegaloviremia consequent to immunosuppressive therapy.

Interferons may become an important adjunct in tissue or organ transplantation, sinced they are the only agents known to combine immunosuppressive and antimicrobial activities. Presumably this is the basis for initial reports that interferons suppress cytomegaloviremia and viruria in transplant recipients treated with immunosuppressive drugs, and that interferons suppress graft-versus-host disease and graft rejection. The herpesviruses (especially Epstein-Barr and cytomegalovirus) are the principal cause of mortality among transplant recipients, who are not merely rendered defenseless by immunosuppressive drugs: it is established that immunosuppressive drugs, and apparently immunosuppression in general, are well-correlated with the induction and dissemination of latent herpes- and papova-viruses (BK, JC and other SV40-like viruses). It is an urgent priority to explore the potential of interferons for simultaneous immunosuppression and antimicrobial effect, as well as their potential in other modes of immunomodulatory therapy. The potential of interferons as immunomodulatory agents may run from transplantation therapy, on one hand, to the management of simple hypersensitive or allergic status, on the other.

Genetics

A new chapter has been opened with the discovery of a gene (Mx^{+}) in mice which is required for expression of the antiviral effect of interferons in a *virus-specific* manner, i.e., against a myxovirus but not against a rhabdovirus. The antiviral effects of interferons are mediated by regulation of the activities of a number of enzymes. Probably, some of these are involved in the antiviral pathways against some, but not other, virus replicative strategies. Possibly, the *Mx* locus illustrates one of these.

The discovery of the *Mx* gene (as well as a half dozen loci controlling interferon *synthesis* serves as a signpost for the future: the applicability of interferons, lymphokines or other "biological response modifiers" may depend on the *host genotype* in a manner quite unlike conventional chemotherapeutic agents. This, in turn, suggests that the responses to interferon therapy could be far more idiosyncratic than is desired, which is just what has been seen in clinical trials thus far.

One suspects that the immunogenetics of the host, and the necessity for such information on a patient-by-patient basis, will play a far greater role if management employing "biological response modifiers" is to succeed. In the worst case, the application of interferons may require knowledge of certain genotypes in a manner analogous to histocompatibility typing. In the best case, application would have been as simple as for classical antibiotics, where idiosyncratic genotype-dependent responses (i.e., allergy) are rare. It seems clear that the application of interferons will not be so simple.

Interferon Inducers

Some mention should be made of the pharmacologic future of interferon inducers. To date, most attention has focused on the use of exogenous interferons, in lieu of inducers of endogenous interferons, since the latter are generally toxic substances whose administration frequently results in a period of hyporesponsiveness to reinduction of interferon. These facts have deterred clinical research with interferon inducers. Nonetheless, research continues which is aimed at finding nontoxic interferon inducers, and studies have suggested that the hyporesponsiveness toward reinduction may be manipulated using other substances. In pharmacologic terms, inducers may become a large and complex subject unrelated to studies with exogenous interferons. Clearly, both approaches are expected to meld when more is known of the biology of the interferon system.

Interferons in Pathogenesis

Considerable evidence now indicates that interferons are important as pathogenic substances. The use of neutralizing anti-interferon immunoglobulins *in vivo* has yielded fascinating evidence concerning (1) the role of interferons in limiting the course, and in *defining the pathologic picture* in viral and neoplastic diseases, and (2) in aggravating the pathogenesis of other diseases, especially those with an immunopathologic component. Properly speaking, the latter refers not only to autoimmune states, but also to malignancies and persistent virus infections.

The roles of interferons in pathogenesis in humans are speculative at present. However, there are two types of reasons for expecting that interferons are pathogenic mediators in humans:

The first of these, in Gresser's words, is "guilt by association": interferons, including unusual (acid-labile) IF-alpha, are present at elevated levels in a large proportion of patients with autoimmune diseases such as lupus, rheumatoid arthritis, scleroderma, systemic vasculitis and Sjogren's syndrome, as well as in patients with acquired immunodeficiency syndrome (AIDS) and, in all probability, other immunopathological disorders.

The more compelling reasons for assuming that interferons are pathogenic mediators arise from the fascinating studies by Gresser and his colleagues in animal models. These studies first discovered that, in strains of mice susceptible to neonatal infection by lymphocytic choriomeningitis (LCM) virus, the fulminant liver necrosis caused by the virus is actually caused by interferon induced by the virus infection. The liver pathology could be caused by pure interferon in the absence of the virus and is prevented during LCM infection by administration of anti-interferon immunoglobulin. It was then found that if sublethal doses of interferon are administered for a few days to suckling mice and rats, immune complex glomerulonephritis occurred reproducibly later in life in animals which were, until then, clinically well. Ultrastructural studies showed that perinatal interferon treatment did, in fact, induce immediate changes in kidney maturation which eventually led to clinical disease and death. In another mouse model for autoimmunity where immune complex kidney disease is present (i.e., the murine model for lupus, NZB/NZW disease), the pathogenic effects of immune complexes and complement deposition in kidneys, lung, vascular and cardiac tissue are accelerated by interferon treatment. Interferon-antibody complexes are among those found deposited in the kidney. *In vitro*, interferons also mediate the formation of "lupus inclusions" in lymphoblastoid cells.

While studies *in vitro* and in animal models may have no direct bearing on clinical studies in humans, they signal the care which will be required in bringing interferons into medical practice.

We are far from a complete understanding of the roles of the endogenous interferon systems. It seems obvious, *a priori*, that such information will be required for rational use of interferons, since situations may be found where endogenous interferons are already present, or are pathogenic. As a corollary, exogenous interferon therapy would seem futile when there are lesions in the efferent arm of the interferon system (hypore-sponsiveness to interferon action). At present, only a small number of studies have explored situations in which the endogenous interferon response may be defective.

A related area concerns the potential of interferons in pediatric use. At least in newborns, who are very susceptible to disseminated virus infections, the IF-alpha and IF-beta responses are already functional. The occurrence of neonatal virus infections may arise from immaturity of the monocyte-macrophage lineage. This, in turn, suggests that interferon therapy may be contraindicated in neonates because it is inhibitory toward the maturation of monocytes and granulocytes.

More generally, the cytostatic activities of interferons may render them dangerous in newborns, and in pregnancy. The marked embryotoxicity of interferons and IF-inducers is well known but unexplained. On one hand, this may lead to a new postconception method of contraception. On the other hand, further study of interferons as possible mediators of teratogenic effects seems warranted. The latter suggestion arises because (1) in humans, production of, and sensitivity to, IF-alpha and IF-beta are mature within two weeks of gestation, (2) interferons are inhibitory toward both growth and cellular motility *in vitro*, and (3) differential growth and cellular motility are key morphogenetic mechanisms. If this reasoning is correct, then interferons could be involved in the teratogenic effects caused by certain virus infections *in utero*, as well as in teratogenic effects of transplacental interferons arising from maternal tissues. In this regard, it seems reasonable to ask whether the dysmorphogenesis characteristic of trisomy-21 (Down's syndrome) is linked to the enhanced sensitivity of trisomy-21 cells to IF-alpha and IF-beta. One might reasonably wonder whether the capability of producing, and responding to, these interferons so early in life does not signal a physiological role for these substances in embryogenesis.

CONCLUSION

In overviewing areas of interferon research, I have attempted to include those observations which will be relevant to the successful use of interferons in medicine. To these can be added a plethora of further observations of fundamental interest, which space does not here permit.

The overriding goal of interferon technology serves as a model for other, newly-discovered, immunomodulatory proteins: to realize the promise of clinical utility which these substances appear to hold.

This goal cannot be achieved without a commitment to the *biology* of interferons which is far greater than that which has already gone into the *technology* of their production by recombinant methods. The substances of which this chapter speaks are many; they are potent; they are pivotal in the maintenance of health and the mechanisms of disease; and their physiological actions are extremely diverse. Without such a commitment, the technology producing interferons will never have commercial significance arising from their successful application in medicine. It would be tragic, indeed, if government and industry fail to recognized that "basic" and "applied" interferon research will be inseparable for many years to come.

Because our knowledge of the biology of interferons is so rudimentary, considerable further research at the animal, *in vitro* and biochemical levels is required in order to realize the promise inherent in these substances. In particular, it seems self-evident that clinical trials in humans cannot be truly informative without concurrent and greatly expanded studies of related diseases in animal models, particularly the mouse.

We are only at the beginning of the journey which will be required to derive widespread utility from interferons. We now have the tools to produce, modify and study these fascinating substances at will. But in view of the multiplicity of interferons and of their actions (to say nothing of new molecules created by recombinant techniques), many years of study remain. One can envision a time when the cycle of classical chemotherapeutic development—modify the molecule and test its activity—will apply to interferons and related cytokines. Equally clearly, our knowledge of the biology and pharmacology of these substances is primitive; in remedy, the power of recombinant technology and nucleic acid sequencing are quite irrelevant. Here, as in other walks of life, the "technology" has already quickly outstripped our ability to employ it rationally.

16

MONOCLONAL ANTIBODIES

Clinical laboratories commonly use one or a combination of techniques for the diagnosis of infectious diseases, including microscopic examination of tissues and exudates, culture methods, and measurement and identification of either antigens in patient specimens or specific antibodies in their sera. However, rarely has one method proven optimal for all situations. For example, to detect certain microorganisms direct microscopic identification is adequate; but for maximum sensitivity, many organisms must be cultured. Culture methods are very sensitive, but are also labor-intensive and time consuming because a positive culture must be confirmed with a second test. In addition, some common pathogens are difficult or impossible to culture.

Immunologic techniques for direct identification of antigens can provide rapid and specific diagnostic methods, but often lack adequate sensitivity, and are dependent on the availability and specificity of the antisera used. Close antigenic relationships exist within many groups of microorganisms, and antisera may cross-react among species or between pathogens and non-pathogens. Serum testing for an antibody response to microorganisms is commonly performed, but the presence of serum antibodies does not always accurately reflect an active infection or immune response to a specific pathogen. With recent developments in monoclonal antibody technology, however, there is the opportunity to reevaluate the use of specific immunologic probes for the diagnosis of microorganisms.

MONOCLONAL ANTIBODY TECHNOLOGY

Based on the pioneering studies of Kohler and Milstein, it is now routinely possible to prepare immortal, cloned cell lines which continuously and reproducibly secrete unique antibody molecules by the chemically mediated fusion of lymphocytes from an immunized animal and a mouse myeloma cell line. These cell lines can be obtained at high frequency, and with appropriate screening techniques hybrid cells can be selected that produce antibodies which critically distinguish antigens in different species or strains of microorganisms. These antibody molecules demonstrate precise specificity, react with uniform avidity, and can be readily purified to homogeneity from either culture fluids or mouse ascites fluid. Monoclonal antibody techniques are now widely practiced, and antibodies of diagnostic potential have been prepared in research laboratories against a battery of viruses, bacteria and parasites.

We have developed three panels of monoclonal antibodies for the detection for the detection of human sexually transmitted infections in our laboratories at Genetic Systems Corporation. One panel reacts specifically with *Chlamydia trachomatis*, another panel reacts with *Neisseria gonorrhoeae*, and the last panel distinguishes herpesvirus types 1 and 2 from each other and from other common microorganisms. In a joint program with Syva Company these antibodies will be used to develop a line of rapid, easy-to-perform diagnostic tests.

Human Sexually Transmitted Infections

With changes in sexual attitudes and behaviour in our society, sexually transmitted diseases have become increasingly commonplace. In the U.S. alone, new infections produced by *Neisseria gonorrhoeae*, *Chlamydia trachomatis* and herpes simplex virus are believed to exceed ten million cases annually. In recent years these infections have been implicated in a wide spectrum of diseases, some of which have such serious complications as systemic infection, infertility, perinatal infection, and neoplasia.

Chlamydia Trachomatis

Chlamydial genital infections are now recognized as a major health problem in the United States, with an incidence estimated at several times that of *Neisseria gonorrhoeae*. *C. trachomatis* is the causative organisms in up to 50 percent of cases of both nongonococcal urethritis in men, and its counterpart, mucopurulent cervicitis in women. Both have increased dramatically over the last decade. Some infections may progress to more severe disease such as endometritis, pelvic inflammatory disease, perihepatitis, cervical dysplasia, or salpingitis in women, and epididymitis in men. In addition, Chlamydia may be transmitted perinatally, resulting in conjunctivitis and/or respiratory infection in the neonate. *C. trachomatis* infection is also associated with a number of conditions including Reiter's syndrome, and in developing countries, lymphogranuloma venereum and endemic trachoma.

Though chlamydial infection may be effectively treated with antibiotics, the inability of most physicians to obtain cultures and the nonspecific patient symptoms these infections produce have made control difficult. In an attempt to increase the availability of a sensitive, specific diagnostic test, we have used a fluorescein-labeled monoclonal antibody in immunofluorescence (IF) tests to significantly improve the sensitivity of the Chlamydia culture assay and to reduce the time needed for diagnosis. We have also used this fluorescein-labeled monoclonal antibody to develop a new, rapid, culture-independent diagnostic test in which extracellular chlamydial organisms are detected in patient specimens.

Monoclonal antibodies to C. trachomatis

A panel of monoclonal antibodies reacting with *C. trachomatis* was prepared by cell fusion. For immunodiagnostic purposes we selected a monoclonal antibody reactive with the principal outer membrane protein (40,000 daltons) on all serovars (immunotypes) of human *C. trachomatis*, but not reactive with *C. psittaci*. The antibody was purified from mouse ascites fluid and conjugated to fluorescein isothiocyanate.

The IF staining method was compared to the more traditional iodine staining method for the detection of *C. trachomatis* in cultures which were inoculated with patient specimens. Urethral (men) or cervical (women) samples were taken from patients with calcium alginate swabs, immediately placed in transport medium, and refrigerated until inoculation (within 24 hr) in quadruplicate onto McCoy cell monolayers grown 96-well microtiter plates. Cultures were incubated at 37° and a well was then stained with either fluorescein-labeled monoclonal antibody after 2 days or iodine after 3 days. The two remaining cultures were then disrupted, inoculated into fresh cell monolayers, and incubated an additional 2 or 3 days before IF or iodine staining.

Table 16.1. Comparison of iodine and IF staining methods for detection of *C. trachomatis* in culture.

Iodine Stain	+	–	+	–
Monoclonal antibody Stain	+	+	–	–
Number of Specimens	417	57	23	2288

At second culture passage concordant results with the two assays were observed in 2705 of the 2785 specimens tested (97 percent).

Staining of specimens with antibody yielded 474 positives (17 percent) compared with 440 positives (16 percent) when stained with the iodine method. Of the 440 specimens that were positive by iodine stain, 417 were also positive in the IF test. If iodine-stained cultures were considered as the reference method (that is, to provide 100 percent "correct" diagnosis), the overall sensitivity for the antibody method was 95 percent. The IF test has a specificity of 98 percent and predictive values of 88 percent and 99 percent for positive and negative specimens, respectively. The percentage of positives were not significantly different with specimens obtained from men or women.

When analyzed at first passage the IF method detected 25 percent more positive specimens than iodine stain. This difference narrowed, however, by the second passage where the IF method detected only 8 percent more positives than the iodine stain. With almost all specimens from first or second passage, the IF method detected considerably more inclusions per monolayer than the iodine method.

In results, 98 percent of all positive specimens were identified with IF staining at the first culture passage at 2 days, whereas with iodine staining, 38 percent of infections were not apparent until the second culture passage at 6 days. Because of the increased sensitivity of the IF technique it may be possible to eliminate the second passage in tissue culture. This would allow reporting of accurate results more rapidly, with an additional decrease in the laboratory workload. Since the IF test takes less time to read, and has a lower proportion of uninterpretable cultures than iodine stain, there can be considerable savings in technician time as well.

Direct diagnosis of chlamydial infection

In an effort to further reduce the time required to diagnose chlamydial infection, we have developed a slide test that can be performed on specimens obtained directly from patients. Slides were stained with monoclonal antibody and scanned at 400X magnification under the fluorescence microscope. Positive specimens were identified by the presence of stained extracellular chlamydial elementary bodies which appeared as individual pinpoints of apple-green fluorescence. With higher magnification (630X) these extracellular chlamydial organisms cold be accurately resolved as a discrete, smooth-edged, evenly fluorescent elementary bodies produced a delicate "starry sky" pattern that could be observed throughout the sample. Although intracellular inclusions could be readily seen in cells infected *in vitro*, they were only rarely seen in direct clinical specimens. This direct slide test was compared for its value in the diagnosis of chlamydial infection with cultures that were stained with iodine or monoclonal antibody. For this purpose a swab containing a urethral or cervical sample was smeared onto a microscope slide. A second swab was then taken and placed in transport medium for subsequent inoculation into culture. The specimen slide was fixed in ethanol for 1 minute and stored at –20°C until stained with fluorescein-labeled monoclonal antibody.

The results of this study indicated comparable levels of sensitivity between the direct slide test and the two culture methods used. For 926 patients, the direct slide test identified a total of 211 positive specimens (23 percent). At first culture passagethe IF stain detected 208 positives (22 percent) while iodine stain detected 184 (20 percent). With second culture passage, the total number of IF positives was essentially the same at 205 (22 percent) while the total number of iodine positives increased to 199 (21 percent). Complete agreement among the three methods was observed with 93 percent of the specimens tested. When cultures stained with iodine were considered the reference method (100 percent "correct" diagnosis), the direct slide test provided a sensitivity of 92 percent and specificity of 96 percent, with predictive values of 87 percent and 98 percent for positive and negative specimens.

Table 16.2. Comparison of culture methods and direct slide test for detection of Chlamydia in adult urogenital specimens.

Diagnostic Test	*Pattern of Reaction*							
Direct Slide Test	+	–	+	+	–	–	–	+
Culture-Antibody Stain	+	–	–	+	+	–	+	–
Culture Iodine Stain	+	–	–	–	+	+	–	+
Males	117	468	13	5	5	5	8	2
Females	65	225	6	2	3	1	0	1
Total	182	693	19	7	8	6	8	3

Quantitative examination of the data demonstrated that the specimens in which discrepant reactions were obtained were derived primarily from patients with low-titered infections. In these, the number of inclusions observed in cultural rarely exceeded 5 per well, while the number of elementary bodies observed in the slide test rarely exceeded 20 per specimen. Since multiple swabs were used to obtain specimens from each patient, it was also considered likely that sampling error was a significant factor in these discrepant reactions.

We have extended our studies utilizing the direct slide test to detect *C. trachomatis* in ocular and respiratory tract specimens obtained from infants. These studies demonstrated a high correlation between results of the direct test and results of IF and iodine culture methods. For 55 specimens taken from the conjunctiva and from the nasopharynx, there were 16 positive specimens, with a complete agreement between direct IF testing and culture. Results of direct IF tests and culture also correlated for 12 of the 13 oropharyngeal specimens (3 positive and 9 negative) taken from infants with conjunctivitis or rhinitis. A single sample which was culture-positive and slide-negative, produced a very low count of 2 inclusions per well.

Collectively, these findings suggest a rapid and simple approach to the diagnosis of chlamydial infection. Specimens are obtained on a sterile calcium alginate swab, smeared on microscope slides, placed in transport medium, and the slides tested by the direct IF assay for the presence of extracellular chlamydial elementary bodies. At least 90 percent of chlamydial infections should be detected by this procedure. At the discretion of the physician, cell cultures may be inoculated with smear-negative specimens, terminated at 48 hours, and stained with the monoclonal antibody to detect intracellular chlamydial inclusion bodies. This latter procedure should detect any low-titered infections that were not apparent by direct testing.

C. trachomatis infections are easily treated with tetracycline or erythromycin. However, lack of available diagnostic testing has made control of infection difficult. Increased use of chlamydial diagnostic techniques would facilitate case detection and contact tracing, and may be expected to have an impact similar to that of increased culture diagnosis on the incidence of gonococcal infections. Use of the direct slide test will permit many laboratories that are not equipped for mammalian cell culture to perform a rapid and simple diagnosis of chlamydial infection.

Neisseria Gonorrhoeae

From 1960-1974 the incidence of gonococcal infection in the U.S. increased at a rate of about 12 percent per year. As a result of increased control measures, its incidence has stabilized since 1975 at approximately 1 million new cases reported annually to the Centers for Disease Control in Atlanta. It is estimated that the true incidence of gonorrhea in the U.S. exceeds 2 million cases yearly. *N. gonorrhoeae* is a cause of cervicitis in women, and can produce urethritis, pharyngitis, and proctitis in both sexes. As with chlamydial infection, gonococcal infection may produce serious sequelae including epididymitis in men, or endometritis and salpingitis in women. A small percentage of infected individuals develop disseminated gonococcal infection with systemic complications. Gonococcal infection has also been implicated as a cause of morbidity during pregnancy, and when perinatally transmitted can cause neonatal conjunctivitis, an important cause of blindness in developing countries.

Detection of gonococcal infection involves either Gram stain and microscopic examination of the patient's specimen, or culture on selective medium and subsequent confirmatory assay. In urethral samples from men, a positive Gram stain has a predictive value of >95 percent when compared with results of culture and requires only a few minutes for performance. Gram stain in women, however, is considerably less sensitive, detecting less than 50 percent of the cervical infections diagnosed by culture. Consequently, culture methods are preferred for diagnosis of gonorrhea in women, even though 48-72 hours are needed for completion.

Monoclonal antibodies to N. gonorrhoeae

We have produced a panel of 16 monoclonal antibodies which react with different epitopes on the principal outer membrane protein (protein I; 34-37,000 daltons) of the gonococcus. When these individual antibodies were used in coagglutination tests, each detected only a subset of the *N. gonorrhoeae* reference strains. Antibodies selected for broader reactivity against gonococci were found to cross-react with other *Neisseria* species, decreasing their diagnostic value. Pooling several monoclonal antibodies against protein I into a defined polyclonal mixture would identify the entire spectrum of *N. gonorrhoeae* strains, without compromising the selective specificity of each antibody.

To select an appropriate antibody mixture, we screened each of the 16 different monoclonal antibodies in coagglutination assays with 719 different isolates of *N. gonorrhoeae*. Each antibody reacted with a characteristic subset of gonococci, resolving two broad, mutually exclusive serological groups of gonococci expressing serologically distinct forms of major outer membrane protein, termed protein IA and IB. Five of the monoclonal antibodies detected determinants on the protein IA molecule. For example, antibody 4-G5 reacted most commonly, identifying 99 percent of the protein IA strains, whereas antibody 4-A12 demonstrated a more restricted range of reaction, detecting only 44 percent of the protein IA strains. Eleven other monoclonal antibodies detected determinants on the protein IB molecule. Antibody 2-H1 reacted most commonly, identifying 93 percent of the protein IB strains. Some of the other antibodies tested, 3-C8, 1-F5, 2-

G2, and 2-D4 reacted with 78, 52, 18, and 17 percent of the protein IB strains respectively.

On the basis of their patterns of reaction, we pooled three of the monoclonal antibodies (4-G5, 2-H1, and 3-C8) and tested this mixture with the 719 isolates. As could be predicted from the results of tests performed with the individual antibodies, the antibody mixture identified 716 (99.6 percent) of the isolates tested. This same antibody mixture, when tested on 18 different *Neisseria* species, reacted exclusively with *N. gonorrhoeae*. Thus, in the case of gonorrhea, the construction of an antibody mixture proved to be a satisfactory method to overcome the limited specificity observed with individual antibodies. We are currently testing the efficacy of this antibody pool to detect gonococci in direct patient samples using an immunofluorescence assay.

Herpesviruses

Herpes simplex virus (HSV) infections are among the most common infections of humans. Once acquired, each infected individual serves as a permanent carrier with life-long episodic recurrences. The spectrum of disease associated with these viruses may range from asymptomatic infection to potentially fatal encephalitis.

Herpesviruses have been traditionally classified into 2 subgroups: HSV type 1 is responsible for recurring facial lesions, while HSV type 2 is responsible for genital infections. However, it is now apparent that this "anatomical" classification is not truly accurate, as recent reports show that 15 percent of primary genital herpes is caused by HSV 1.

HSV 1 infections are widespread, with almost 80 percent of adults demonstrating anti-HSV 1 antibodies in their sera. HSV 2 infections are becoming increasingly more common, with approximately 300,000 new cases of genital herpes occurring each year. These accrue upon an estimated base of 5 to 10 million individuals who are already infected.

There is now an increased need for the rapid and specific laboratory diagnosis of HSV infection since the advent of anti-viral chemotherapy for both mucocutaneous and visceral HSV infections. In addition, differences in the natural history of genital HSV 1 and HSV 2 infections and the sensitivity of the two viruses to certain anti-viral agents have made laboratory HSV typing important for both prognosis and therapy.

Clinical identification of HSV infection has been generally performed by isolation of virus in tissue culture. For the purposes of immunological typing, antisera produced in animals have been used in immunofluorescence or neutralization tests to distinguish HSV 1 from HSV 2. However, due to the close genetic relationship between the two viruses, conventional antisera obtained from rabbits, goats, or convalescent patients contain mostly cross-mostly cross-reacting antibodies, and definitive serotyping is often difficult to perform. More recently, serotyping procedures have been improved with the use of monoclonal antibodies that specifically react with either HSV 1 or HSV 2.

Monoclonal antibodies to herpesviruses

For the purposes of identification and typing of herpes simplex viruses, we have developed a panel of four monoclonal antibodies that unambiguously distinguish HSV 1 from HSV 2. Monoclonal antibody 3-G11 reacts with the HSV 1 specific 80,000/120,000 dalton glycoprotein (gC) complex, antibody 6-A6 reacts with an HSV 2-specific protein of 140,000 daltons, antibody 6-E12 reacts with an HSV 2 specific protein of 55,000 daltons, and antibody 6-H11 reacts with an HSV 2 protein of 38,000 daltons.

Culture diagnosis of herpesviruses

With this panel, we and our collaborators have now identified and serotyped over 550 different clinical isolates of HSV.

Table 16.3. Comparison of methods of typing of Herpes simplex viruses.

	No. Virus Isolates			
Test Method	*HSV1 alone*	*HSV2 alone*	*Indeterminate*	*Mixed HSV 1/HSV 2*
Rabbit	11	70	41	NA
Monoclonal antibodies	34	83	0	5
Restriction endonuclease	34	83	0	5

In this study, 122 HSV isolates from 107 patients were typed by three independent methods (immunoperoxidase-labeling with type-specific rabbit antisera, IF with monoclonal antibodies, and restriction endonuclease analysis of viral DNA). Results of monoclonal antibody typing demonstrated a 100 percent concordance with restriction endonuclease analysis of viral DNA. In 117 isolates HSV was unambiguously typed as either HSV 1 or HSV 2. In five different isolates from three patients, the monoclonal antibodies typed a mixed infection of HSV 1 and HSV 2, which was confirmed by plaque purification of viruses from the mixture and by restriction endonuclease analysis of both the virus mixture and the plaque purified viruses. In contrast, commercial antisera prepared in rabbits could definitively type only 66 percent of the 122 isolates; the remaining 34 percent of isolates yielded indeterminate antigen patterns from which a definitive identification could not be made.

Direct diagnosis of herpesviruses

In addition to their utility in culture systems, the monoclonal antibodies also provided sufficient specificity to enable diagnosis and typing of HSV directly on primary clinical specimens. For this purpose, cells were obtained from herpes lesions by scraping with swabs. They were then smeared onto microscope slides for IF tests with fluorescein-labeled monoclonal antibodies against HSV 1 or HSV 2. In each test a duplicate sample of the specimen was inoculated into cell cultures. IF tests were performed on specimens (from oral, genital, mucocutaneous, and ocular sites) obtained from 59 patients with clinically suspected HSV and 43 control patients with unsuspected HSV infection.

With specimens obtained from suspected herpes lesions, 48 of the 54 specimens (88 percent) in which HSV was isolated in tissue culture had HSV antigens detected in IF tests. The monoclonal antibodies did not detect HSV antigens in any of the 43 specimens obtained from the control population. The antibodies could identify and serotype viruses in clinical specimens from a variety of anatomic sites demonstrating genital HSV-1 infection, oral HSV-2 infection, and HSV-2 infection from diverse mucacutanious sites.

It should be noted that the antibodies failed to detect HSV antigens in six direct specimens from patients with genital HSV infection, while culture methods revealed infectious viruses. Subsequent testing of the six isolates (obtained by the culture method) demonstrated that each reacted with either 6-G11 or the three monoclonal antibodies against HSV 2. This indicated that the "false negative" results with the direct clinical specimens were based on insensitivity of the IF method, rather than a failure of the antibodies to recognize viral antigens. The inability of the antibody method to detect all HSV infections may be related to the quantity of cells in the specimens. Although samples containing less than 20 cells were rejected from the IF analysis, it may be necessary to increase the number of cells required for sample acceptance in order to provide a greater chance of observing virus-infected cells, and thus increase test sensitivity.

In ten specimens, HSV antigens were detected by the monoclonal antibodies, but virus was not isolated by tissue culture. Previous investigations of IF methods for HSV antigens have also demonstrated viral antigens in clinical specimens in the absence of infectious virus. This may be due to partial genomic expression, intermittent viral shedding, presence of neutralizing virus, or loss of infectivity during transport.

The monoclonal antibodies described here appear to have sufficient specificity to provide both diagnostic and serotyping information on direct clinical specimens. Because prior studies using IF tests with homotypic antisera from animals have shown unreliability in serotyping clinical isolates, the monoclonal antibodies appear to offer an additional advantage over conventional anti-HSV reagents.

Conclusions

To summarize, monoclonal antibodies are already being used for the routine diagnosis of human sexually transmitted diseases. For both culture and direct tests, monoclonal antibodies showed patterns of specificity and reproducibility that exceeded those available with conventionally prepared antisera. The direct tests for these organisms required less than an hour to perform, representing a major advancement in the diagnosis of disease which previously required 2 to 6 days of culture followed by confirmatory testing. Furthermore, rapid, differential diagnosis of infection will now be possible. Since some sexually transmitted diseases may be co-transmitted and share similar clinical manifestations (i.e., gonorrhea and Chlamydia in cervicitis or urethritis; syphilis, chancroid, or herpes in genital ulcers) it will be possible to differentiate a single from multiple infection by simultaneous testing of direct samples with the appropriate monoclonal antibody reagents.

17

USE OF MONOCLONAL ANTIBODIES

The presence of *Salmonella* in foods and feeds continues to be a major worldwide problem. In 1980, 30,004 isolations of *Salmonella* from humans were reported to the Centers for Disease Control in the United States. This figure represents only a small fraction (only about 1%) of the total cases; the remainder are unreported. It also does not include non-human cases of *Salmonella*.

Surveillance of salmonellae by conventional cultural procedures is both costly and time consuming. This procedure requires a minimum of 96 hours for a negative test and costs approximately Ω11 per assay. Food companies must store their products until a negative test is obtained, thereby incurring additional costs. Therefore, a simple, accurate, rapid method to detect salmonellae has been an elusive dream of the food industry for some time.

Rapid procedures to detect salmonellae in foods have basically followed two approaches: (1) those based on the biochemical or growth characteristics of the organism, and (2) those based on their serological, or antigenic characteristics. Methods that rely on the first criterion are usually not very rapid and may lack the necessary specificity. Immunological methods, on the other hand, are sensitive and rapid, but have suffered from a high percentage of false positive reactions. These false positives are due in part to the non-specific nature of the polyclonal antibodies used in these techniques.

The fluorescent antibody technique was one of the first immunological methods utilized for the detection first of salmonellae.

This procedure was reported by Thomason *et al.* in 1957. An excellent summary of the history of the development of the antibody conjugates used in this technique has been published by Thomason. Early antisera varied in antibody content and were not specific enough to detect the various serotypes of salmonellae found in foods and feeds. Eventually, the reagents were improved and a polyvalent OH antiserum was made available in 1972. With the availability of standardized antisera, the FA techniques was approved as official final action by the Association of Official Analytical Chemists (AOAC) in 1976. The method was also recognized by the Food and Drug Administration (FDA) and the American Public Health Association.

Even with the improved antisera, however, the FA technique still is a victim of false positives. Cross reacting antibodies remain in the antisera and cannot be removed without also losing the reactivity for salmonella. The use of pure IgG antibodies for the assay is an improvement, but does not totally eliminate the problem. Typically, the false positive rate for this assay is between 1 and 9 percent.

Although it is a more rapid procedure (45-50 hrs), the fluorescent antibody technique has failed to gain a significant foothold in the laboratories of the food industry. In addition to problems with false positives reactions, there are other disadvantages. All positive results must be confirmed using conventional cultural testing. Microscopic equipment needed to perform the assay is expensive and a specific space must be set aside for FA analysis. Personnel must be well trained to recognize the positive fluorescent reactions and fatigue is a problem in reading large numbers of samples by the FA technique. Although monoclonal antibodies could improve some aspects of this test, some negative factors would still remain, this test could not be automated, and technician fatigue would still be a consideration.

In 1969, Sperber and Deibel reported on an accelerated procedure to detect *Salmonella* in dried foods and feeds called enrichment serology (ES). Pre-enrichment and selective enrichment are performed as in the conventional culture procedure, followed by a 6 hr elective enrichment in M-broth. Direct serological testing, using pooled H antisera, is performed on a portion of the M broth culture, with test results available 50 hr after test initiation. Although the ES procedure correlated 100% with the cultural procedure in Sperber and Deibel's study, Mohr, *et al.*, found a false positive rate of 13%-15% with the ES procedure, and a 2% false negative rate. The false positive results were again due to the cross-reactivity of the polyclonal Spicer-Edwards pooled H antisera. The false negative results were attributed to insufficient outgrowth in the M broth enrichment. Although rapid and inexpensive, the high false positive rate has precluded the wide acceptance of this assay on a routine basis. In addition, *S. agona* is a species not detectable by the pooled antisera suggested by Sperber and Deibel. Once a rare serotype, *S. agona* in 1980 ranked 6th in the serotypes isolated from human sources.

The most recent methodology to employ an antibody system for the detection of *Salmonella* in food products is the enzyme immunoassay (EIA). An EIA was developed in 1977 by Krysinski and Heimsch, but the antibodies which they used were not pure. Minnich improved this procedure by purifying the IgG portion of the antiserum and eliminating the IgM fraction, thus reducing the amount of cross reactivity of the antibody preparation. Swaminathan and Ayres developed a direct EIA for detecting salmonellae in foods, but the procedure required that all preparations be fixed to glass slides and observed microscopically. As in fluorescent antibody procedures, use of microscopy has drawbacks and is not widely acceptable. In 1982, Minnich, *et al.* developed in indirect EIA procedure, but still the antibody preparation used was not ideal.

The initial study using a monoclonal antibody in an enzyme immunoassay for *Salmonella* was reported by Robison, *et al.* in 1983. The monoclonal MOPC 467, an IgA myeloma, was able to specifically detect salmonella organisms even in mixed culture, with no cross-reactivity to other enterics. Unfortunately, the antibody was incapable of detecting all of the *Salmonella* serotypes tested.

The MOPC 467 myeloma protein was isolated by Potter in 1970 and was later found to be specific for a flagellar determinant found on several *Salmonella* strains. The only organisms that cross-reacted with the antibody were *Pastuerella pneumotropica* (now Yersinia) and *Herellea vaginicola*. Those organisms would not be expected to be present in routine food analyses. Potter concluded that the determinant recognized by MOPC 467 is not an H antigen, since strains exhibiting a wide variety of H antigens are recognized by the monoclonal antibody. Also, certain strains of *Salmonella* are bound with greater avidity than others, i.e., *S. typhimurium* and *S. anatum* give weaker reactions than do *S. milwaukee* and *S. mississippi*. The only known serotypes of Salmonella not recoganized by MOPC 467 include *S. paratyphi A*, *B*, and *C*, *S. typhi*, *S. kirkee*, *S. tennessee*, and *S. newington*. *Arizona hinshawii*, also known as *S. arizonae*, is however recognized by the monoclonal antibody.

At the very heart of a successful EIA lies the specificity and affinity of the antibody for the antigen to be detected. The antibody is then coupled to an enzyme by one of several methods. We chose coupling alkaline phosphatase to MOPC 467 by the gluteraldehyde method, as reported by Voller, *et al.* The antigens to be detected are affixed to a solid phase and we used 96 well microtiter plates. The labelled antibodies are incubated with the attached antigen for one hour, and then thoroughly washed, such that the only enzyme present in the well must be attached to the antibody which must be attached to the antigen. A colourless substrate which can be cleaved to a colourimetric compound is then added to the well; the formation of a colour is indicative of the presence of antigen. We used para-nitrophenyol phosphate (Sigma), which is cleaved by alkaline phosphatase to yield a yellow colour. This can then be read in a spectrophotometer at 405-415 nm, and the optical density is recorded.

Because the MOPC 467 antibody attaches to a flagellar antigen, we chose to heat the bacterial samples (30 minutes, 100°C) and attach the proteins of the supernatant to the microtiter plate. This allows for the attachment of more relevant antigens/well, as well as allowing a very tight binding of the relevant antigen to the plastic. Heating of the organisms is not necessary to detect the antigen, however, as Smith and Jones described an EIA using MOPC 467 on unheated samples. Their assay did not attempt to attach the whole organisms to a microtiter plate, but rather used the organisms themselves as the solid phase immunoadsorbent in a competition-type assay.

We prepared heat extracts from over 100 strains of salmonellae and other enteric organisms and attached these to the plates. After performing the assay, we interpreted a positive reaction to be one which gave an optical density (OD) reading of greater than 0.1 above a heat extract of *E. coli*. Table 17.1 lists the strains which were tested and whether they were positive or negative in this assay. As is readily seen, MOPC 467 appears to detect a large majority of the *Salmonella* strains which are important in the food industry, while showing no false positive reactions.

In order to determine the sensitivity of the assay, we diluted the heated supernatants of a few strains to correspond with known numbers of organisms. Table 17.2 shows that all strains tested were detected at a level of 10^6 *Salmonella*/ml, and some strains could be detected at much lower levels. What this means, in practical terms, is that, assuming a 30 minute division time for *Salmonella*, a 6 hour culture which had

Table 17. 1. *Salmonella* species tested.

+ adelaide	+ hartford	+ oranienburg
+ agona	+ havana	+ orion
+ albany	+ heidelberg	+ oslo
+ allandale	+ illinois	+ panama
+ anatum	+ indiana	– paratyphi A
+ archavaleta	+ infantis	– paratyphi B, bio java
+ arizonae	+ inverness	+ poomona
+ bareilly	+ javiana	+ poona
+ bere	+ johannesburg	– potsdam
+ berta	+ kentucky	– pullorum
+ blockley	+ kingston	+ raus
+ braedenburg	– kirkee	+ reading
+ braenderup	+ korovi	+rio grande
+ bredeney	+ kottbus	+ rostock
+ calabar	+ krefeld	+ rubislaw
+ cerro	+ lexington	+ saint paul
+ champaign	+ litchfield	+ san diego
+ chester	+ livingstone	+ schwarzengrund
+ cholerae-suis	+ london	+ senftenburg
+ cholerae-suis bio kunzendorf	+ manhattan	+ siegburg
+ cubana	+ mbandaka	+ simsbug
+ derby	+ meleagridis	+ stanleyville
+ drypool	+ minnesota	+ tallahassee
+ dublin	+ mississippi	– tennessee
+ dusseldorf	+ montevideo	+ thompson
+ eastbourne	+monschaui	– typhi (9 strains)
+ enteritidis	+ muenchen	+ typhimurium
+ essen	+ muenster	+ typisuis
+ florida	+ new brunswick	+ uganda
– gallinarum	– newington	+ urbana
+ gaminara	+ newport	+ virchow
+ give	+ norwich	+ waycross
+ glostrup	+ ohio	+ weltevreden
+ hadar	+ onderstepoort	+ westlaco
		+ worthington
		+ zongo

Other Genera tested

– Citrobacter (4 strains, 3 species)

– Edwardsiella (1 strain)

- Enterobacter (4 strains, 3 species)
- Escherichia (8 strains)
- Klebsiella (5 strains, 2 species)
- Proteus (2 species)
- Providencia (3 species, 2 strains)
- Pseudomonas (1 strain)
- Serratia (4 strains, 2 species)
- shigella (2 species)

been inoculated with only 500 organisms would be detectable with this assay. Certainly pre-enrichment cultures should yield greater than 500 *Salmonella*/ml, which would indicate that detectable numbers of *Salmonella* should easily be obtained within two working days under optimal conditions.

Table 17.2. Sensitivity of the EIA with MOPC 467.

	Absorbance (410 nm) at indicated cell concentration (per ml)			
Organism	10^8	10^7	10^6	10^5
E. coli	0.228	0.220	0.144	0.142
S. koroni	2.232	2.232	1.695	0.262
S. infantis	2.068	1.872	0.336	0.196
S. minnesota	1.475	1.448	0.273	0.079
S. anatum	2.210	2.232	0.771	0.245

Since it is imperative that an EIA for the food industry be able to identify as many *Salmonella* strains as possible, we decided to produce monoclonal antibodies against the strains which are not detectable by MOPC 467. We thus immunized BALB/c mice with treated organisms of *S. typhi*, the paratyphis, and other strains of *Salmonella*. Fusions of their spleen cells were performed using a variant of the Sp 2/0 cell line, a myeloma line widely used in cell fusions. This line produces no immunoglobulin of its own and divides approximately every 24 hours. The technique of Lerner was used, which is a modification of the technique of Milstein and Kohler.

Approximately 3 weeks following fusion, the supernatants of the resulting hybridomas were tested for reactivity against *Salmonella* antigens of different strains. Although several wells produced positive results, many of these lost the ability to form antibody after subsequent expansion and cloning. Some produced antibodies which cross-reacted with other enterics, and some produced antibodies of only low titer

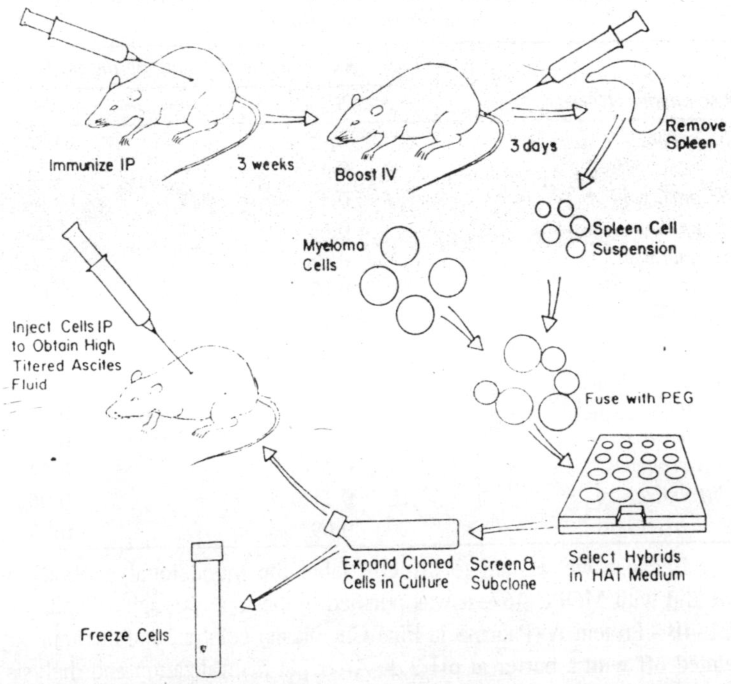

Fig. 17.1. Preparation of Hybridomas.

toward *Salmonella* antigens. One of them, labelled 6H4, produced an antibody which showed strong reactivity against many different strains of *Salmonella* while showing no reactivity to other enterics. This was cloned and expanded, and the supernatant was used in an indirect EIA.

As can be seen, the 6H4 antibody identifies every known strain which the MOPC 467 does not detect. In addition, it is seen that it identifies several strains which MOPC 467 also detects, and it does not attach to heat extracts of other enterics. Because these two antibodies (MOPC 467 and 6H4) have differential binding properties, it is assumed that they do not identify the same antigenic determinant. The indirect EIA utilizes a second immunoglobulin to detect binding, an enzyme labelled goat anti-mouse immunoglobulin.

In order to determine the isotype of the 6H4 monoclonal antibody, the supernatant was attached to the microtiter wells, and labelled anti-isotype antibodies (anti IgM, IgG1, IgG 2b, IgG3) were then incubated in the wells. Results indicate that 6H4 is an IgG 2b. It is now growing as ascites tumors in BALB/c mice, and the cell lien has been frozen for later use.

Table 17.3. Comparison of the specificities of the different monoclonal antibodies.

Organism (10⁷/ml)	Absorbance (10 nm) of antibody MOPC 467	6H4
S. koroni	0.95	0.76
S. typhi	0.13	1.11
S. paratyphi A	0.13	0.98
S. paratyphi B	0.09	1.28
S. tennessee	0.08	1.10
S. newington	0.07	0.67
S. kirkee	0.12	1.14
S. infantis	0.25	1.05
S. typhimurium	0.46	0.95
E. coli	0.11	0.03
P. villgaris	0.09	0.05
E. aerogenes	0.15	0.04
Citrobacter sp.	0.11	0.06
S. marscesens	0.09	0.03

Because we wished to directly label the monoclonal antibody as we did with MOPC 467, it was purified by passage through a Sepharose CL-4B—Protein A (Pharmacia Fine Chemicals) column, and the antibody eluted off with a buffer at pH 2.3. After pH neutralization and dialysis, the antibody was labelled with alkaline phosphatase as before. These antibodies were tested against MOPC 467 non-reactive strains of *Salmonella*. Antibody concentrations as low at 5 μg/ml can detect all of the *Salmonella* strains tested.

Table 17.4. Ability of monoclonal 6H4 to bind to *Salmonella* strains which are not detected by MOPC 467.

Organism	Absorbance (410 nm) using 6 H4 Ascites (μg/ml) 100	50	10	5	2	0.5
E. coil	0.04	0.04	0.03	0.04	0.01	0.03
S. typhi	2.0+	2.0+	1.12	0.60	0.33	0.11
S. paratyphi A	2.0+	1.82	0.80	0.36	0.14	0.05
S. paratyphi B	2.0+	2.0+	1.02	0.58	0.28	0.10
S. newington	1.12	0.72	0.48	0.23	0.10	0.04
S. tennessee	2.0+	2.0+	1.00	0.47	0.26	0.10
S. potsdam	2.0+	1.68	0.72	0.35	0.12	0.05
S. kirkee	1.42	1.00	0.64	0.28	0.11	0.06
S. gallinarum	1.85	1.32	0.85	0.41	0.18	0.08
S. pullorum	0.18	0.16	0.12	0.11	0.11	0.08

These antibodies were then tested for sensitivity, again using heat extracts of different salmonellae such that they correlated to known numbers of organisms. Although showing differential sensitivities, all organisms were detected at 10^6/ml or less. These results, when compared with the results of labelled MOPC 467 would indicate that the combination of MOPC 467 and 6H4 could detect almost all strains of *Salmonella* at a concentration of 10^6/ml or less.

Table 17.5. Sensitivity of monoclonal 6H4 (10 μg/ml) toward different concentrations of *Salmonella*.

	Absorbance (410 nm) at indicated cell concentration (per ml)			
Organism	*10^8*	*10^7*	*10^6*	*10^5*
E. coli	0.08	0.08	0.06	0.03
S. typhi	2.0+	2.0+	1.18	0.46
S. paratyphi A	2.0+	1.42	0.54	0.08
S. tennessee	2.0+	1.46	0.71	0.24
S. newington	1.78	1.06	0.32	0.06
S. infantis	1.86	0.98	0.34	0.16
S. cerro	2.0+	2.0+	1.12	0.41

We therefore mixed phosphatase-labelled MOPC 467 and 6H4, each at 20 μg/ml, and added this mixture to microtiter plates having heat extracts of *Salmonella* attached to them. This allowed detection of every *Salmonella* strain, with the exception of S. pullorum. We do not feel that this negative result is necessarily related to the lack of flagella, as S. Gallinarum, a non-flagellated species, is detected by 6H4.

The development of this EIA should prove to be a tremendous asset to the food industry. It creates an inexpensive, rapid, automated, easy to perform assay which is virtually 100% accurate.

Although we think that a two day assay could be performed, it would require (a) the inoculation of at least 500 Salmonella to be inoculated from the pre-enrichment culture, and (b) extensive work at the end of the second day. This protocol gives more time than is essential at every step, and virtually assures accurate results. Even the inoculation of 1 *Salmonella* organism would theoretically be detectable by the assay following the recommended 2 hour culture period of M broth.

Aside from the obvious advantage of this assay over the culture methods, which include time and ease, the big advantage is cost savings.

There is no need for selective media, no large technician time requirements for plating the organisms and media preparation, and no problems with disposal. Incubator space is also saved, as the assay is performed at room temperature. Products can be released sooner, which eliminates some need for storage space. It shows no cross-reactivity (with the exception of *Herellea vaginicola* and Yersinia species), and is very sensitive. No expensive equipment is necessary, no special training is required, and there is no need for a special license or problems with disposal, as with radioimmunoassays. In addition, the results are objective, not subjective, as with fluorescent antibody assays.

Why this assay using monoclonal antibodies was not developed sooner for the food industry is a mystery. The specificity of the MOPC 467 antibody was reported in 1970, and the solid phase EIA was created in 1971. As far back as 1972, Carlsson, *et al.* were using an EIA for determining antibody responses to Salmonella 0 antigens. This was mainly to be used in the medical field.

It is hoped that, in addition to the food industry, this assay will be utilized in the diagnosis of human infection. In these cases, since 10^6 organisms would be expected to be present in stool samples, no cultural procedure would be necessary, and results could be obtained within 2 hours of obtaining the sample. Medical treatment could then be started at least 24 hours before a positive culture could be obtained.

In summary, we have created a rapid, inexpensive, and specific EIA using monoclonal antibodies for the detection of *Salmonella*. As far as we know, the only limitation to virtual 100% effectiveness is that the concentration of organisms be 10^6/ml or greater. The results can be read visually or by a spectrophotometer/EIA plate reader.

18

MONOCLONAL ANTIBODIES TO BACTERIAL ANTIGENS

The antigenic characterization of bacteria for epidemiological and taxonomic purposes and the generation of vaccines against the pathogenic species is as old as the discipline itself. As in other areas of cell biology and in microbiology, the impact of hybridoma technology has been such that vast and new dimensions in antigenic characterization have opened in ways not conceivable ten years ago. Unlike its sister branch of microbiology, virology, few comprehensive reviews exist which address themselves specifically to the use of monoclonal antibodies (MCA) as applied to the study of bacteria *per se*. This chapter is intended to fill this deletion, as well as to draw attention, where possible, to overlapping areas in other biological disciplines where this powerful probe has enabled a precision in antigenic and functional analysis not possible before.

Many reviews on monoclonal antibodies and lymphocyte hybridomas have been published which are valuable and informative exposes of the properties and applications of these unique biological tools. They address both the potential use as well as any limitations or restrictiveness of the same. Methodology has also been dealt with in considerable detail and eloquence. In addition to the overall general applicability in biology of these references, reviews on virology have special relevance to the use of MCAs to bacterial antigens.

Definitions and Terminology

The following terms are used in the "language" of antigenic characterization introduced first by Jerne and emphasized by Yewdell and Gerhard, their increasingly common usage makes it appropriate to include them in this chapter.

Paratope — refers to the structure of the antibody combining site.

Epitope — refers to the antigenic structure which contributes to the free energy of interaction with a given paratope.

Epitype — refers to the set of epitopes with which a given paratope cross-reacts.

Antigenic site — refers to the individual or groups of epitopes that are topologically non-overlapping.

In this chapter, MCA is chosen as an abbreviation for monoclonal antibody(s) and is used throughout.

Overview of Monoclonal Antibodies: Their Limitations and Their Applications in Antigenic Analyses

Some pertinent points which are relevant to an overview of the use of monoclonals as a tool for antigenic characterization are the following. Firstly, the cells from the spleen are the most frequently used source of mouse B cells from which one expects the elaboration of antibody. They are used almost exclusively in monoclonal antibody production. However, other organs can serve equally as well. The spectra of antibody isotypes and specificities can be unique and distinct depending on the organ source and mode of immunization used to obtain parental primed lymphocytes for a putative hybridoma. The mucosal response is quite different in terms of cell source and distribution than is the systemic. In some cases where a single surviving clone is used as the source of a monoclonal antibody, it is important to remember that it represents only one cell of the millions of the animal's repertoire of cells. An antibody derived from another clone may behave differently when used for antigenic characterization.

As a corrollary to this, it is additionally important to consider the number of individuals used as antigen-primed lymphocyte donors, and the immunization protocol in order to generate a diverse panel of hybridoma antibodies which will be representative of the antigenic repertoire of the microorganism. Both the physical and biological characteristics of the antibody are a function of the selected host's response.

The mode of antigen preparation, route and method of immunization must be clearly defined for a given objective in the use of MCA for antigenic analyses since the single cell, random distribution of fusion products dissects the whole response into its individual components. A biased selection of the individual components as in only certain hoped for clonal responses can be made. The bias can be compounded through a deliberate selection in antigen preparation. Hence, important antigenic determinants can be overlooked, or certain responses may not occur.

It is important to have a specific knowledge of the immune status of the experimental animal used to generate the reagents in question in order to interpret meaningfully the outcome and specificity of a given reagent or antibody of choice. An awareness of shared or unshared determinants is essential in interpreting unexpected positive or negative test results. The term "monospecific" may be misleading in the sense that a given paratope has the ability to interact, although with different affinity, with related epitopes. It is equally important to be mindful that cross-reacting antigens need not share common functions or genetic origins. At the same time, the knowledge that the preferential response of the mouse to polymorphic determinants both in prokaryotic cells and in eukaryotic cells is essential in interpreting unexpected positive or negative findings when tested against a panel of MCAs. A panel of monoclonal antibodies may discern only a portion of the total antigenicity of a microorganism. There is no guarantee that all relevant specificities will be obtained.

Technical Points of Consideration with Application to Procedures

Since procedures have been well outlined by others the following will not be a detailed discussion of the same. Rather, points of consideration will be drawn to the attention of the reader which are intended to supplement existing information in these references. Other chapters in this same volume doubtless draw attention to important points as well. A generalized recommendation for those embarking on monoclonal antibodies for the first time is to visit a laboratory where the "system" is set up and working, or to participate in a course. The NIH and other organizations have offered these from time to time. There is no substitute for a "hands on" experience under qualified tutorship in a field which is still delicate and complex, and in which many of the cellular events are still unknown. The details involved from the initial culture of parental myeloma lines to a good exponential growth phase, through a critical fusion step with appropriately primed B cells, then through several months of cloning and subcloning of hybrids and elimination of variants to a point of satisfactory genomic stability are too numerous for the uninitiated to cope with from simply reading the literature. The most fastidious of bacteria are not as capricious as the laboratory-created eukaryotic tetraploids (hybridomas), each of which has its own random, but unique chromosome complement which may be more or less stable.

Selection of parental cell line

Myeloma parent

The user must have a complete and thorough knowledge of the characteristics of the myeloma line (most frequently, but not necessarily, of murine origin) chosen as the fusion partner. These include growth and cultural characteristics which will be imparted in whole or in part to a putative hybrid cell. If unique drug resistance markers are included in the choice of the myeloma line, any restrictions imposed by the same must be carefully considered. An example is cumulative toxicity with time. Such is the resistance of MOPC 315.43, the line used in our laboratories, to 1mM ouabain. Even though the hybrid cells are selected efficiently through the elimination of all unfused spleen cells, they are routinely placed in medium without ouabain once they have been identified as clones of choice to avoid their kill through cumulative toxicity.

Yet another consideration is the secretor status of the myeloma parent. Hybridomas which express immunoglobulin genes from the myeloma parent as well as those from the normal B cell partner do not secrete a structurally homogeneous population of antibodies. If the myeloma-derived (heavy or light 1g chains) show distinct specificities for the antigen under study in addition to the B cell-derived 1g chains, the interpretation of the antigenic analysis is very complex. Coexpression of chains from both parents *within* a single hybrid cell leads to the expression of mixed molecular species.

The myeloma cell culture should be in exponential growth phase at the time of fusion. Increased granulation and self-agglutination, observable under the phase microscope are morphological features which are indicative of a stage of senescence beyond exponentially growing cells in some lines. It may also be an indicator of toxicity or other poor culture conditions.

Normal B cell parent

Most often this is derived from the spleen which can produce desired isotypes under carefully chosen immunization schemes. Fusions have also been done with gut-associated lymphoid tissue. The important thing is to use B cells when their production of antibody can be demonstrated. Often more antibody is better, although for purposes of definition of the host response or delineation of the host's immune status, specific stages may be desirable, as for example, the specificity unique to the early response. In short, it is important to know when, under what conditions of concentration, mode of antigen administration and preparation, and to what extent the response occurs in order to tailor the harvest of activated B cells for experimental objectives. This in addition to overall strategy, is eloquently described by Galfré and Milstein. It can be accomplished by assaying the polyclonal antibody in serum during immunization schedules of which there should be more than one with several animals. Observations made with individual monoclonal antibodies should always be evaluated in the context of the overall antigenicity as delineated by conventional antisera.

Assay systems

In addition to a good, efficient assay system for screening clones together with supplementary assay systems as required, the following are important points of consideration.

Specificity

Specificity as referred to above, and demonstrated in the early 1970s by Cunningham and Pilarski occurs to the highest degree early in the immune response and becomes more diverse with time and continued priming. It is not necessarily accompanied by the highest overall antibody response. If a measurement of specificity is important to the experimental protocol, a panel of antigen controls which appropriately measures this, must be utilized early in the characterization of the product of any clone.

pH and temperature conditions

Mosmann, et al. have shown that the apparent specificity of a given MCA to polymorphic determinants can be altered depending on the physical conditions of the assay system. This would be expected from classical protein chemistry but has not received great attention in reviews of monoclonal assays. It can be the suspect source of clone supernates which fail to react with time when stored at refrigerator temperature for continuous use and the pH indicator of the medium shows an obvious change. Conversely, it can be used to advantage in making an antibody "more" specific by changing pH.

Immune status of the animal used as a source of B cells

All animals show a background antibody level ("natural antibody") which is most likely derived from physiological exposure to antigens of similar epitype. Because the normal bacterial flora of all animals is extensive under conventional living conditions, this is of special consideration for bacteriologists. Intentional immunization of an animal which already possesses background antibody does not produce a primary response, but rather a secondary or tertiary response. Specificity and antibody titre are the most immediate parameters affected. However, a thorough knowledge of the immune status can influence the design of both the mode of immunization and antigen preparation to induce a desired response. A cautionary note is advised in the context of an *a priori* selection of monomorphic determinants which can bias an interpretation.

The following discussion is an attempt to illustrate the range of antigenic analyses available to the bacteriologist through hybridoma technology and monoclonal antibodies. The above comments are reflected in the literature to a greater or lesser extent. In some cases assay systems used in MCA-antigenic characterization are designed to detect biological functions. In other cases, assays define physical or chemical characteristics of the antigen. The former has the more immediate priority interest in clinical or therapeutic potential. Some antibodies described are human MCAs derived through the use of B cells selected from the existing repertoire of antibody-producing cells, without deliberate priming.

Monoclonal Antibodies to whole Bacteria and Source Antigens

Monoclonal Antibodies Against Bacterial Capsules

Haemophilus influenzae

Hunter, et al. described the use of spleen cells from a splenectomized patient which were fused, using polyethylene glycol, with a mutant human myeloma cell line deficient in hypoxanthine-guanine-phosphoribosyl transferase (i.e., sensitive to HAT medium). The report described the characterization of the antibacterial activity of an IgG secreting, biologically active antibody against the type b capsular polysaccharide of *Haemophilus influenzae*. Enzyme immunoassay was used to screen fusion products against the capsular polysaccharide even though the patient from whom the spleen was taken (a thrombocytopenic purpura unresponsive to steroid therapy) had no known infection of, or vaccination against *H. influenzae* type b. The product of a single clone thus screened was assessed for biologic activity *in vitro* using a human neutrophil-mediated opsonophagocytic assay in which 5, 10, and 20 μg of antibody were incubated with 5×10^6 colony forming units of *H. influenzae* and 10^6 neutrophils and rabbit complement. Subsequent plating of the suspensions from microtitre wells was used to determine the percent of kill of the original bacterial suspension relative to controls with complement only, neutrophils only, or the two combined without antibody. The activity of the antibody was inhibited with purified *H. influenzae* capsular polysaccharide, but not with type XIV pneumococcal antigen (a control of unrelated specificity). No killing effect was observed when Type III group B streptococci were substituted for live *H. influenzae*, but was greater than 90 percent for four additional *H. influenzae* type b isolates. The antibody was protective in the suckling rat model (67 percent survival of 2-day old rats 5 days after infection when given 200 μg of the monoclonal antibody as opposed to 15 percent survival of the control group given saline in place of antibody).

Similar experiments were done with mouse MCA of the IgM class, capable of fixing complement *in vitro* in a bactericidal assay. It was also effective in reducing both the level and incidence of bacteremia when given to previously infected infant rats.

Streptococcus pneumoniae

Type III pneumococcal capsular polysaccharide has been used to generate MCAs, although not with the objective of defining the capsular material functionally, but rather through its utilization as a T-independent antigen in characterizing an X-linked immunodeficiency in mice. This is described in greater detail below under the definition and characterization of the host response to bacteria.

Escherichia coli and Neisseria meningitidis group B capsules

Certain *E. coli* capsular antigens (K antigens) are associated with invasive disease. In an attempt to analyze factors affecting bacterial colonization and virulence, Söderström, et al., prepared monoclonal antibodies to K, pilus, and somatic antigens. Included in the representation were the capsular polysaccharides of *E. coli* K13, K1, K20, and K23. These were conjugated to proteins such as bovine serum albumin (e.g., K13-BSA). MCA against the K1 capsular polysaccharides protected neonatal rats against bacteremia. MCA to K1 and K13 protected mice against death. All antibodies to capsular antigens efficiently opsonized *E. coli* in opsonic assays which also occurred with antibody against O antigens, but not with antibody against pilus antigens.

The K1 polysaccharide is immunochemically identical to the group B meningococcal polysaccharide and a monoclonal antibody of murine origin of the IgM class to group B polysaccharide was shown to kill

K1-positive strains in the presence of human complement in an opsonophagocytic assay. It was used, in addition to other preparations, in an evaluation of immunotherapeutic approaches. The same antibody was used as a homogeneous, specific reagent for both meningococcal group B and *Escherichia coli* K1 polysaccharides in comparing the bactericidal activities with complement from different sources. Differences were evident between complement from humans and rabbits (greater in rabbits in this case). It was felt that the potential existed for characterizing more closely the terminal complement complex, C5-9, particularly as it relates to individuals deficient in the complex who suffer from repeated meningococcal infections.

Type-Specific, Group-Specific Polysaccharides and Surface-Exposed M Protein of Streptococci

Group specificity of streptococci is determined by the "C substance" or cell wall polysaccharide. Further immunologically distinguishable polysaccharide structures include the "S substance" (type subdivisions in group B), as well as capsular polysaccharides like those in *Streptococcus pneumoniae*. All are exposed to a greater or lesser extent on the surface of the bacterium. The chemical structure and composition is fairly well characterized and correlated with immunological distinctiveness. In addition, proteins on the surface or "within" the capsule such as the M protein of group A streptococci and R protein of group B streptococci are also immunogenic.

Formalinized whole cells of type II and type III group B streptococci were used by Polin to immunize Balb/c mice. Fusion products were tested for binding activity to bacterial cell pellets using ^{125}I anti-mouse Fab and an enzyme immunoassay. Six out of 10 antibodies specific for type III antigen detected with the binding assay and subjected to further analysis also agglutinated type III group B streptococci. None of the antibodies against type II or group B streptococci agglutinated type II cells. Immunodiffusion against immunoglobulin class-specific reagents was used to determine the class of antibodies.

Of particular interest in this set of experiments was the observation that these antibodies did not cross-react with type XIV *S. pneumoniae*. The core antigen of type III group B streptococci and the capsular polysaccharide of type XIV *S. pneumoniae* are immunochemically similar insofar as a trisaccharide backbone of galactose, fucoseamine and glucose in a molar ratio of 2:1:1 exists in both. However, type III native antigen contains a terminal sialic acid attached to each galactose end group and does not cross-react with type XIV pneumococcal antisera. Hence, the lack of cross-reactivity of the above cited monoclonals may be suggestive of formation against the type III native structure. It may also lend supportive evidence to a generalized preferential response, to this epitope of the type III antigen in infants recovering from invasive disease caused by type III group B streptococci.

In other experiments by Ruch and Smith, two selected mouse monoclonal antibodies against the group-specific carbohydrate antigen (as opposed to the type-specific considered to be the outer most of the cell wall carbohydrates) were intensively characterized for their value as diagnostic reagents. One in particular, precipitated both purified group B antigen and antigen present in autoclaved or enzyme extracts (common diagnostic methods for streptococcal grouping) of group B streptococci. It was specific in gel diffusion reactions and counter-immuno-electrophoresis. When coated on latex particles it compared well with commercially available polyclonal reagents. Insofar as a reagent highly specific for the group antigen offered the potential for use in latex agglutination (simple and quick in diagnostic laboratories) of all five serotypes, its development was of particular interest. An aspect noted was that the most abundant carbohydrate of the group B antigen, rhamnose, did not inhibit either binding (radioimmune assay) to whole cells or precipitation assays of either of the two antibodies selected. This is in contrast to rabbit polyclonal antibody from which the inference that rhamnose is the immunodominant determinant was made. This apparent immunodominance reopens a question as to whether there is a relationship between animal species, or whether the two antibodies selected happened to react with other epitopes, or whether unknown complex factors yet to be, identified may lead to a recharacterization of an immune response.

A mouse monoclonal antibody reacting specifically with the terminal N-acetylglucoseamine of streptococcal group A carbohydrate has been reported to react identically in Lancefield precipitin test with commercial reagents and, more specifically in immunofluorescence.

In experiments designed to define more precisely the protective determinant of M protein exposed on the surface of group A virulant streptococci, it was found that neither the ability to precipitate a purified polypeptide fragment of M protein preparation nor the ELISA titre of hybridoma antibodies correlated with opsonic, bactericial or protective activity. Only some of a set of hybridoma antibodies prepared were opsonic against homologous type streptococci. When injected into mice, the opsonic antibodies also had protective activity against challenge infections with the homologous, but not heterologous serotype of bacteria. One of the antibodies which reacted in extremely high dilutions in the ELISA and was not opsonic, also failed to react with intact streptococci, suggesting that the antigenic determinant for that MCA was not available for binding or was "buried" in the native state of the M protein. However, alternative explanations for this behaviour of antibody to a complex bacterial cell surface macromolecule which renders the microorganisms resistant to phagocytosis in the non-immune host could include: the determinant to which the antibody was directed was not involved in the antiphagocytic activity of the M protein; the internal location of the determinant on the streptococcal surface may not have been recognized by complement or phagocytic cells; the determinant may have been a "new" or "modified" entity created by the preparation procedure insofar as the purified polypeptide was used as the immunogen. The overall thesis of this work argues that since MCAs, each directed to a single distinct immunodeterminant can be generated, protective immunity may be directed to anyone of several distinct antigenic determinants of M protein exposed on the bacterial cell surface. However, a single antigenic determinant could evoke type-specific protective immunity if presented appropriately to the host.

Outer Membrane Antigens of Gram-Negative Bacteria

Except for the capsule, the outer membrane of gram-negative bacteria is, for all intents and purposes, the first structure of the intact bacterium to come in contact with the host immune system and its ancillary mechanisms. Together with the capsular polysaccharides, both protein and lipopolysaccharide components of the outer membrane have been the objects of endeavors to classify, type, and determine the functional role of the membrane in interactions with the host. Their immunogenicity in many cases has invited the immunochemical and immunobiological routes for characterization.

Haemophilus influenzae

Hansen et al., had previously found that infant rats recovering from systemic disease produced by *Haemophilus influenzae* type b (Hib) mounted an antibody response directed against outer membrane proteins. Moreover, young human infants recovering from meningitis also responded to outer membrane proteins, whereas their developing immune system does not effectively mount an antibody response to the capsular polysaccharide. Protective studies using the infant rat model in which animals were injected with 80 μg of MCA protein and challenged six hours later with 100 to 200 Hib colony forming units showed decreased quantifiable bacteremia 24 hours later. The same protection was not observed against heterologous Hib strains which did not contain the identical 39 K determinant. The antibody also eliminated established infection. Although the antibody did not provide the advantage of an overall type protection as those described above against capsular antigens, it did characterize an antigen which is immunogenic for a population vulnerable to this particular organism.

Neisseria gonorrhoea

Gonococci have been shown to exhibit colony variation in terms of their opacity or transparency properties during the course of an infection and it is thought that antigenic variation during an episode of infection may be partly responsible for the inefficient clearing mechanisms of the host. An MCA, generated through the immunization of mice with whole gonococcal cells, reacted with whole cells of transparent and intermediate opaque phenotypes using a radioimmune assay. Lack of the specific antigen was correlated with a deeply opaque colony type. When outer membrane proteins were electrophoresed (SDS-PAGE) there was a protein present in variants which expressed the antigen and absent in those lacking the antigen. It was acknowledged that the availability of a target antigen which could be affected by the presence or absence of the variable protein was a possibility equally as likely as that the variable protein was the antigenic determinant itself. Nevertheless, the point was made that variable expression of antigens was amenable to study with an MCA probe.

An extension of outer membrane protein characterization of gonococci was made by Tam, et al., in establishing a panel of 16 phenotypically stable, independently cloned hybrid cell lines secreting characteristically different MCAs which reacted with a unique antigenic determinant on protein I of the outer membrane complex of *Neisseria gonorrhoeae*. In antibody binding, immunofluorescence and coagglutination, the antibodies reacted with a restricted group of *N. gonorrhoeae* strains. They also did not react with other *Neisseria*. The MCAs demonstrated serological relationships between strains of reference serotypes paralleling those of conventional polyvalent antisera. The MCAs recognized at least three different epitopes and permitted the subdivision of the 34 reference serotypes into two independent serological groups on the basis of sharing of common peptides (tryptic peptide maps). Hence, protein I could be divided into IA and lB. Some of the monoclonals recognized subsets of IA and IB. As a panel, these MCAs demonstrated their utility in precise serotyping for epidemiological studies, allowed identification of serogroups, and were amenable to fluorescein labelling for immediate and more sensitive detection and speciation of gonococci in cervical exudates.

Some of the same authors have extended the work into systematic diagnostic utility. This report devotes itself to immunodiagnosis of sexually-transmitted diseases with MCA and, hence, includes herpes virus. However, the emphasis in this chapter is on bacterial antigens such that the pertinent points respecting *N. gonorrhoeae* and the family *Chlamydiaceae* with the one genus *Chlamydia* will be referred to.

The authors adapted replicate-plating techniques from bacterial genetics for testing antibody specificities by transferring culture fluids from 96-well microtest plates to similar test plates containing different antigens adsorbed to the wells. ^{125}I-labelled protein A was added and the wells examined by autoradiography. A two-tiered selection system was used in which antibodies reacting exclusively with organisms within a single family or genus were selected first. Secondly, subsets of these antibodies were selected for their ability to identify as many members of a family or genus as possible. A point of difference to be noted in organisms examined is that in the case of *Chlamydia*, which have a unique intracellular developmental cycle, a single MCA against an outer membrane protein could be used to distinguish all members of *C. trachomatis* without cross-reacting with the one other species in the genus, *C. psittaci*. With *N. gonorrhoeae*, each of the MCAs against the principal outer membrane protein detected only a subset of the reference strains. Antibodies with broader reactivities cross-reacted with other *Neisseria* species. Hence, for practical diagnostic purposes, MCA against protein I were pooled to obtain a defined polyclonal diagnostic reagent. These reagents allowed specific, direct testing of clinical specimens with fluorescein conjugates in a time frame of 15 to 20 minutes, a major advance in the diagnosis of infections which may require three to six days of culture. This comprehensive paper also details many technical aspects in hybridoma technology as it relates to infectious agents with prospects for future immunodiagnostics.

Neisseria meningitidis, group B, and Pseudomonas aeruginosa

In our own laboratory we used semi-purified outer membrane extracts for MCA production to some serotype antigens of *Neisseria meningitidis* group B after finding that the number of hybridoma clones of interest could be markedly increased over using whole cells. A previously described polymorphism was confirmed through the generation of MCA, but specificity of the host response appeared markedly dependent on previous experience with the immunizing antigen. This previous experience would have been minimal or nonexistent in the case of *N. meningitidis* where a high specificity of the majority of clones to the immunizing antigen was observed (*N. meningilidis* is not known to colonize the mouse and is not pathogenic for the mouse). It should be fairly considerable to an ubiquitous organism such as *Pseudomonas* where, indeed, great diversity of reaction of individual MCAs was observed. In these experiments, as in those of others, cross-reactivity could very well have been manifested by the antibody being directed to a common epitope or core structure.

Leptospira and Vibro cholorae

Monoclonal antibodies directed toward the lipopolysaccharide components of gram-negative bacterial outer membranes have often been utilized in serotyping or grouping of the organisms as opposed to the characterization of the pharmacological/biological activities of bacterial endotoxin (LPS). Therefore, two papers dealing with *Leptospira* and *Vibrio cholorae* are described in this section as opposed to a subtopic of Products and Specialized Macromolecules of Bacteria.

Species of *Leptospira* are subgrouped into the equivalent of serotypes or groups on the basis of LPS antigens. These are referred to as serovars and have a name designation. Eighteen hybridoma cell lines secreting MCA to *Leptospira interogans* serovar *kremastos* were classified into ten distinct groups on the basis of microscopic agglutination with other serovars. Five anti- *Leptospira interrogans* var *canicola* MCAs were divided into three distinct groups on the same basis.

A single cell line producing monoclonal antibodies against the core region of *Vibrio cholorae* lipopolysaccharide was characterized with the view of its diagnostic usefulness. Since its activity in the ELISA was inhibited by the two main serotypes, Ogawa and Inaba, which have an anitgen designated A in common, it was inferred that the antibody was directed to the A determinant. All *V. cholorae* strains tested in coagglutination (MCA absorbed to protein A-containing *Staphylococci*) were agglutinated regardless of serotype. Hence, the inference the antibody was to the core region. LPS from unrelated bacteria did not inhibit coagglutination with the MCA.

Legionella pneumophila

MCA prepared to whole bacterial cells of *Legionella pneumophila*, serogroup 1 are treated in this discussion as though they were outer membrane components, purely on an arbitrary basis. However, since capsular material is normally a polysaccharide, and these surface-exposed epitopes had some characteristics of proteins, the organizational allotment for purposes of a review seems to be within reason. Neither of these structures have been definitively characterized in *Legionella*. Three MCAs of IgM class were selected in this case. They did not cross-react with other known serogroups of *L. pneumophila* nor with unrelated bacterial species in an indirect immunofluorescent-antibody test. They agglutinated sero-group 1 organisms of *L. pneumophila*, making them useful rapid diagnostic reagents. It was not determined, in this case, whether the three MCAs reacted with identical epitopes or not.

Other workers have been able to generate MCA to *Legionella pneumophila*, also by immunization with whole cells, which are reactive with antigens not protein in nature (insensitive to protease), and thus far appear similar to LPS. They react specifically with serogroup 1, by immunoblotting of SDS-PAGE gels. IgG MCA capable of co-precipitating with protein A has been used to isolate the antigen on a protein A-sepharose column. The antigen contains neutral sugars, amino acids, and an amino sugar, and has strong reactivity in limulus lysate assay supporting a probable LPS-like structure.

Other structures

Pili

To date, the known report on MCA against pili is that of Soderstrom, et al., referred to above. Pili of *E. coli* were purified and used for immunizing mice as one of several structures in an attempt to characterize factors affecting colonization and virulence in *E. coli*. Type 1 pili are associated with a capacity to adhere to mucous and certain epithelial cells. In contrast to MCA against the capsular and somatic antigens of *E. coli*, MCA against pili were ineffective in opsonic capacity. Nine of 11 showed cross-reactivity when tested against five different pilus preparations. This is not entirely surprising in light of the fact that mannose inhibited most of the reactions, indicating a defined antigenic epitope. Those not inhibited by mannose were, therefore, evidence of additional epitopes. The MCA against pilus antigens agglutinated whole bacterial cells and inhibited other receptor-mediated agglutination such as that of guinea pig erythrocytes and human sperm by pilus preparations. This lends support to the involvement of pili in mucosal attachment and the possibility of antibody protection at the mucosal surface.

The use of complex extracts or homogenates of whole cells, and taxonomic characterization with MCA to cell walls

Although the use of multicomponent extracts has some drawbacks insofar as the antigenicity of these may not necessarily reflect their antigenicity according to their anatomic arrangement in the bacterium, their use can be justified for the following reasons. Antigens with great functional capacity but present in small quantities in the whole bacterium can be quantitatively or qualitatively enhanced as immunogens when antibody to them might be either diluted or masked in the polyclonal response. As long as the nature of the antigen is understood, an increased number of hybrid clones of interest can often be generated for MCA characterization. When used in a test system, the antibody reactions to them can be more precisely controlled. Further, analysis of their characteristics within the whole bacterium can often be accomplished with precision once MCAs to them have been made.

Homogenates are more complex and suffer from the disadvantage of antigenic changes during the procedure in the same manner as extracts. However, in some cases they have also proven to be one of few, if not the only viable route of antigenic characterization.

With the advent of MCA, the *Mycobacteria* have been more succinctly characterized than ever before through the use of sonicates, and supernatants obtained after passing bacteria through a bacterial press with or without subsequent extraction or purification of components. Ivanyi, et al. used *Mycobacterium leprae* isolated from infected armadillo livers, killed with cobalt irradiation, sonicated, and ultracentrifuged as an antigen source. Four soluble antigens in the supernatant fraction thus obtained were identified with 12 murine MCAs. One, a protein antigen of molecular weight of 12,000 reacted uniquely with one MCA only. This MCA was not found to react with any of 20 other *Mycobacteria* tested, making it a likely candidate for diagnostic application. Two antigens were found to be shared by several other mycobacteria, one of which was polysaccharide in nature as evidenced by protease resistance and supported by the IgM and IgG, class of the reactive MCA. Finally, one other antigen reacted with four of the 12 antibodies which themselves showed marginal cross-reactivity with three other species of mycobacteria. Increased sonication in preparation of the antigen mixture raised the concentration of one antigen, suggesting a cell-wall source. Radioimmunoassay and immunoblots from polyacrylamide gel electrophoresis were the assay systems. This particular report extends work done only slightly earlier by Gillis and

Buchanan. Since MCAs specific for *M. leprae* were produced as well as MCAs with a range of specificities, the possibility of competition antibody-binding experiments for identification and diagnosis of infected individuals at early disease stage exists.

Such, in fact, was the technical approach by Hewitt et al. Competitive inhibition by human sera of the binding of ^{125}I-labelled murine MCA to a solid-phase bound bacterial press supernatant of *M. tuberculosis* identified 71 percent of 41 patients with smear-positive pulmonary tuberculosis. This eliminated the need for elaborate purification of antigen and was comparable in sensitivity to previous serological methods. The MCAs used were shown previously as distinguishing between different strains of *M. bovis* and *M. tuberculosis*.

Taxonomic characterization of the methanogenic bacteria, a group which has received limited attention in antigenic definition, included inhibition-blocking experiments with compounds of known composition and structure. Two dissimilar antigenic determinants expressed by the unique pseudomurein of the cell wall were recognized by MCA produced to different organisms within the family *Methanobacteriaceae*, namely; an epitope involving N-acteyl-glucoseamine (part of the backbone structure), and an epitope involving γ-Glu-Ala of the peptide portion of the peptidoglycan. Multiple testing of a panel of MCAs against one of two strains of *Methanococcaceae* indicated recognition of species epitopes as well as markers of higher taxa.

Monoclonal Antibodies to Products and Specialized Macromolecules of Bacteria

Exotoxins

Clostridia

Cl. tetani

Members of the genus *Clostridia* produce potent exotoxins. Two known reports involving hybrid cells making antibody against *Clostridial exotoxins* include a hybrid of a human B lymphocyte and mouse myeloma cell reported by Gigliotti and Insel. This relatively stable hybrid (cultured for 13 months) produced MCA, bound the B fragment of tetanus toxin as well as intact toxin, and protected mice against tetanus.

Cl. botulinum

A more detailed use of MCAs in exotoxin characterization was through the use of four murine hybridoma cell lines cloned from a single fusion against type C_1 toxin produced by *Cl. botulinum* type C strain Stockholm. The reactions of the four MCAs produced by the four individual clones were tested against type C_1 toxin and type D toxins produced by three different strains of *Cl. botulinum*, one of which was the same strain from which the C_1 toxin was derived. Three of the MCAs reacted with the C_1 toxin only in the ELISA and one of these three neutralized the toxin when injected into mice simultaneously with the toxin. The fourth antibody reacted with type C_1 toxin and with type D toxin from one other strain. It neutralized both toxin types equally well, suggesting a common epitope on the toxin molecules. However, the manner of neutralization of the two neutralizing antibodies was different, implying different mechanisms. At high concentrations of antibody the singly reactive antibody was most efficient in neutralization, but its effectiveness dropped abruptly below a critical concentration (30 μg). Animals injected with the doubly reactive antibody (one or the other toxin, but could be compared to the singly reactive antibody only with homologous toxin) showed prolonged survival (i.e., time to death was extended) with concentrations of antibody below 30 μg. This doubly reactive antibody was also effective in blocking the binding of toxin (both types) to synaptosomes as opposed to the singly reactive antibody. The exact structures which served as epitopes on the toxin molecule(s) were not identified.

Corynebacterium diphtheriae

Diptheria toxin (M.W. 62,000) has two fragments designtated A and B. The COOH-terminal region of fragment B is required for binding to the cell surface receptors. After binding, the NH_2-terminal of fragment A enters the cytoplasm where it blocks protein synthesis. Fifteen MCA were prepared to diphtheria toxin. One of the 14 reacting with fragment B inhibited binding of toxin to cultured cells (Vero cells, with L-cells used additionally as less toxin-sensitive). A second MCA blocked a step in toxin entry into the cell following binding to the cell. The other MCAs were ranked intermediate between these two. Protein synthesis by the cells in culture was measured using ^{3}H-leucine labelling following: (a) incubation of antibody with toxin which was then added to the cell culture, and (b) incubation of the cultured cells and toxin, then adding antibody. The abilities of the antibodies to neutralize toxin and prevent entry were ranked by rates of protein synthesis. Moreover, binding of toxin to L-cells was not inhibited by the same antibody which efficiently blocked binding to Vero cells. Hence, it was inferred that different receptors mediate intoxication in different cell lines.

Cholera and Escherichia coli enterotoxins

Cholera enterotoxin is an oligomer of molecular weight of approximately 84,000, also composed of A and B subunits. Subunit A is further separated into A_1 and A_2 fragments. A_1 fragment is an enzyme with ADP-ribosyl transferase activity, which, when acting on the eukaryotic membrane-bound guanosine triphosphatase (a regulatory component of adenylate cylase activity) results in cylase activity. Subsequent increases in cytoplasmic cyclic AMP induce ion influx changes in the intestinal epithelium which result in massive fluid loss, the hallmark of the diarrheal disease caused by *Vibrio cholorae*. Subunit B is an aggregate of five identical polypeptides, each of M.W. 11,500. Some *Escherichia coli* strains produce a cholera-like diarrheal disease.

Robb et al. reported on the isolation and characterization of seven hybrid cell lines producing antibody to subunits of cholera enterotoxin. Five of the seven clones produced antibody to subunit A; two produced antibody to subunit B and efficiently neutralized the biological activity of cholera toxin *in vitro* in a mouse lymphosarcoma assay. They were specific for cholera toxin only. In contrast to polyclonal antisera, anti-subunit A MCAs exhibited small but detectable neutralizing capacity. Western blot analysis showed that anti-subunit A MCAs also reacted with subunit A of both porcine and human heat-labile enterotoxins of *Escherichia coli* giving a preliminary characterization of immuno-conserved regions of these enterotoxins.

In a larger panel of MCAs prepared to cholera enterotoxin, cholera-specific MCAs, as well as MCAs completely and partially cross-reactive with *E. coli* heatlabile enterotoxin subunit A were detected. It should be noted that the immunization regimens were different in the two experiments. Included in the former were separate injections of subunit A while Lindholm et al. used two hyper-immunization regimens of complete toxin. In so doing, 61 clones out of 70 produced antibody (IgG) to subunit B with a similar distribution of cross-reactivity as those to subunit A, above. MCAs to subunit A did not have neutralizing capacity in a rabbit skin test. Those anti-subunit B MCAs with the greatest cross-reactivity with *E. coli* enterotoxin also had the strongest neutralizing capacity. Moreover, although the amino acid sequence homology is close in the two B subunits, only one third of the total MCAs cross-reacted suggesting marked conformational differences.

Endotoxins

The approach to the characterization of bacterial endotoxins with MCA has not been fundamentally different from that of exotoxins. Insofar as endotoxins of gram-negative bacteria, commonly referred to as LPS (for lipopolysaccharide) is an intergral part of the outer membrane, there are some differences in technical approaches. Either whole bacteria or preparations of LPS might be used as the immunizing antigen or as the test antigen.

LPS of Escherichia coli

Hiernaux et al. investigated idiotypy of single clonal products (i.e., monoclonal antibodies) reacting with LPS of *E. coli* 0113 during the course of primary and secondary responses. Pertinent to the characterization of the LPS itself the following findings are of interest. All seven MCAs investigated (three different isotypes, IgM, IgG_2b, and IgG_3 represented) reacted in passive hemagglutination tests with homologous LPS and with that of *Neisseria lactamica* which shares the epitopes of *E. coli* 0113 polysaccharide chain. Three LPS preparations out of the 18 from different species of gram-negative bacteria reacted with three MCAs. That is, *Klebsiella pneumoniae* LPS reacted with one MCA, two *Pseudomonas aeruginosa* LPS reacted with two other MCAs. Since these reactions were inhibited by native protoplasmic polysaccharide (extracted from the protoplasmic fraction, and free of lipid A from *E. coli* 0113), it was inferred that the epitopes recognized were on the polysaccharide chain of *E. coli* 0113 LPS.

Regions of Salmonella LPS

The generation of murine MCA specific for various regions of *Salmonella* LPS has also been reported. Immunizing bacterial strains included smooth as well as rough strains of different core deletions. The MCAs derived were partially protective against intravenous challenge of mice with 100 LD_{50} of smooth, virulent *S. typhimurium*.

Endotoxin of Bacillus thuringiensis

An unusual endotoxin, the d-endotoxin of *Bacillus thuringiensis* var *thuringiensis* is a proteinacious crystal produced by the bacteria as a parasporal inclusion during sporulation. The parasporal bodies consist of dimeric subunits with molecular weight of about 230,000 held together by disulfide bonds. They are referred to as protoxins. The toxin itself, which is insecticidal, is activated enzymatically by the gut juice of insect larvae *in vivo* as well as by limited proteolysis by trypsin. Two molecular species of the toxins produced sequentially are of lower molecular weight and differ significantly from the protoxin in amino acid composition. Hence, the molecular mode of action and the active site of the d-endotoxin were good candidates for MCA characterization of structure-function relationships. Eight cloned hybrid cell lines produced MCA of five kinds directed against four different antigenic sites. They reacted with the protoxin, and the antigens were, therefore, already present and accessible in the protoxin. Two MCAs bound to two unrelated epitopes and did not compete with any other MCA in competitive antibody binding assays. Three other MCAs inhibited binding of others to different degrees. As in some of the characterizations described above, the *in vitro* binding test did not correlate with the bioassay in toxin neutralization. It was not determined whether an active site or a receptor site for the insect gut epithelium was blocked by the MCAs.

Specialized Macromolecules

Two functional proteins of *Escherichia coli* have undergone studies of topology and structure-function relationships through the use of MCAs. One protein is the lac carrier protein; the other is the recA protein of *Escherichia coli*.

LAC carrier protein of E. coli

The lac carrier protein was purified and reconstituted into proteoliposomes, and the sample for immunization of mice was taken directly from a DEAE-Sepharose column in the last purification step. It was dialyzed against water, lyophilized and reconstituted in phosphate-buffered saline. Five sable clones reacted with the purified protein immobilized on nitrocellulose, in immunoblotting and blocked transport in right-side-out membrane vesicles. One of the five antibodies which showed pronounced blocking of active transport of lactose in the vesicles did not, however, affect the generation of a proton electrochemical gradient, nor did it alter the ability of the vesicles containing the lac carrier to bind p-nitrophenyl-α-D-galactopyranoside. Inhibition of transport activity required a concentration of antibody proportional to the amount of lac carrier in the membrane (three-fold molar excess for 50 percent inhibition). The inhibition also occurred within seconds suggesting that the epitope was accessible on the surface of the membrane.

recA protein of E. coli

The *E. coli* recA protein is small (352 amino acids and a M.W. of 37,800) but complex, with a number of catalytic functions and biological roles.

A population of hybridoma antibodies consisting of one or two species of IgG_{2b} were tested for their inhibition of four DNA - (or oligonucleotide) dependent enzymes: DNA strand annealing, DNA strand assimilation, cleavage of lambda repressor by recA protein, and DNA-dependent ATPase. All were inhibited or reduced by the more or less single antibody-species in proportion to the concentration of antibody. Although it was not surprising that this occurred in an interdependent enzyme series, nevertheless, some fundamental inferences could be drawn. One of these was that the size of the intact IgG molecule (M. W. = 150,000) relative to that of the recA protein made it probable that some of the inhibitory effects, such as reduction of ATPase reaction, were due to steric interference versus a direct blocking of the active site. Dissection of whether the antibody interfered with cofactor binding, caused allosteric change or interfered with oligomer formation was not done.

Microbial enzymes

The analysis of microbial enzymes with MCA has ranged through the testing of recognition of different epitopes on the β_2-subunit of *E. coli* tryptophan synthase, the generation of MCA and their characterization of binding specificity and effects on *E. coli* glutamine synthetase activity, structure-function analysis of lactose permease and β-D-galactosidase of *E. coli*, and innovatively, the detection of *Mycoplasma arthritidis* argenine deiminase in mammalian tissues as a marker of infection.

Special membrane and permease purification techniques to enrich permease were used to overcome previous impairment of immunological methods. These difficulties have been encountered both in preparing pure permease and the low antigenicity of the same. Detailed descriptions for the ultimate derivation and testing of an MCA definitively directed at a permease antigenic determinant exposed on the membrane, but not vital for transport were given.

Activating (increase of enzyme activity) and protecting (prevention of heat denaturation of normal enzyme) antibodies for normal β-D-galactosidase were previously demonstrable in polyclonal antisera. MCAs prepared through the fusion of mouse spleen cells with the mouse myeloma, NS-1, demonstrated the additional presence of inactivating (decrease in activity of normal enzyme and demonstrated polyclonally only with defective enzymes) and null antibodies (able to bind enzyme without inhibiting the other three functions). Since the apparent limitation in the polyclonal response was demonstrated in both rabbits and mice, the possibility of differential species response was eliminated, thereby focusing on the resolution available through the use of a large panel of MCAs.

Species-specific MCA to arginine deiminase of *Mycoplasma arthritidis* which did not react with the same enzyme from *Mycoplasma hominis* or *Mycoplasma pulmonis* was a precise differential diagnostic reagent for detecting infected cells by indirect immunofluorescence. Although the enzyme may be found in other mycoplasmas and some other bacterial species, it is not a normal constituent of mammalian tissues, rendering an MCA application of value in rapid identification and diagnosis.

A representative example of the many types of ongoing work in laboratories everywhere on microbial macromolecules is the utilization of one of the clones generated by us, the MCA of which we termed "pan-reactive" to *Pseudomonas aeruginosa*. It reacted with all serotype strains tested. Since that time the MCA has been shown to be directed toward the porin F protein, a structural protein in the outer membrane of *Pseudomonas aeruginosa*. Cloning of the *Pseudomonas* genes into *E. coli* was associated with production of a cytoplasmic precursor molecule of higher molecular weight than the porin protein itself. Since ^{125}I-labelled MCA reacts with this molecule as indicated by Western blot, the possibility of analyzing the kinetics of assembly of the protein into the bacterial outer membrane, at least to some degree, exists.

Other MCA have been generated to *Pseudomonas aeruginosa* outer membrane antigens. In addition to those reacting with 0 antigen serotypes, one was specific for the major outer membrane lipoprotein H2 which is shared by other *Pseudumonas species*.

Definition and Characterization of the Host Response to Bacteria with Monoclonal Antibodies

B Cell Response

Since the normal activated B cell is the one parent in the hybrid cell line produced, the B cell response is logically a parameter of the host's immune capacity which can be defined on a cellular basis with hybridoma technology. There is no reason to think that fusion does not occur randomly. A good sampling of fusion products should be representative of the original population of host B cells. Ideally, the myeloma parent should be a non-secretor or should not synthesize light or heavy chains which can form mixed immunoglobulin molecules with the Ig chains of the normal B cell parent. Since this line does not synthesize any heavy chain, and synthesizes an altered lambda chain which it cannot secrete, any hybrid cells derived express the Ig chain characteristics of the B cell parent only.

This line allowed us to assess the early immune response of the mouse to extracts of outer membranes of two genera of gram-negative bacteria which are non-pathogellic for the mouse. *Neisseria meningitidis* is not known to colonize the mouse and can be induced to produce disease only under conditions of high, non-physiological concentrations of iron. *Pseudomonas aeruginosa* does not normally produce disease in the mouse but does colonize it from time to time, and the mouse is exposed to it physiologically. We found that out of a total of 475 clones producing MCA to various prototype strains in group B, *N. meningitidis*, as well as five strains isolated from human carriers, 80 percent were uniquely specific for the immunizing antigen. Twenty percent reacted with more than one serotype. The majority of these reacted with only one or two other serotype antigens in addition to the immunizing antigen. Only after a longer period of exposure to the antigen did this increase to a maximum of eight serotype antigens per MCA under the experimental conditions.

The response to *Pseudomonas aeruginosa* indicated only 19.8 percent of 126 total MCAs uniquely specific for the immunizing antigen. Sixty-seven percent reacted with more than one serotype antigen and were referred to as "public" to indicate the similarity to polymorphic allelic systems in eukaryotic cells. "Panreactive," or "species-specific" MCAs which had not been detected at all in the *Neisseria* system made up 12.6 percent. Together with the presence of readily detectable antibody in mouse serum, it was inferred that previous experience with the antigen as in a secondary or tertiary response, allowed the detection of antigens (panreactive) to which there is no preferred response as in the early response to polymorphic determinants.

Many bacterial antigens could fall into this category. Hence, it is helpful to know both the immune status of the animal used for immunization, as well as any preferential responses peculiar to the animal.

In the same vein, most of the serotypes in *N. meningitidis*, group B have been characterized with hyperimmune rabbit serum. In our work, an initial characterization with rabbit polyclonal antibody of carrier strains of *N. meningitidis* showed the following serotype pattern in isolates from a single subject at different times:

week 7 of observation:	serotype 2, 7
week 8 of observation:	serotype 2
week 11 of observaton:	non-typable.

It should be noted that the rabbit antisera used had been prepared to prototype strains within *N. meningitidis*, group B, and could not, therefore identify other unique antigenic components which did not fall into known serotype groupings. Since the above observation indicated an apparent loss with time of carriage of antigenic components in the outer membranes of *N. meningitidis*, the experiments were embarked upon. Murine MCA generated to each of the above isolates in these experiments indicated a predominance of a unique antigen common to the immunizing strains with a random distribution of a minority of MCAs reacting with known serotype antigens including serotype 2 (note that the isolate from week 11 was non-typable with rabbit antisera). At no time were there any MCAs to the subgroup 7 of serotype 2 detected, as in week 7. Moreover, on reexamination of all fusion experiments with mice immunized to prototype strains which included serotype 2 and its subgroups 2, 7, and 2, 10, no clones out of a total of 130 produced antibody unique to an epitope which could be assigned as 7 or 10. Seemingly the mouse, as used in these experiments, did not respond to these epitopes when measured early in the response. All clones detected as being of interest in initial screening in fusions involving immunization with *N. meningitidis* serotype 2 only, were highly specific with no additional cross-reactions. Another fusion done nearly three years later resulted in 54 clones positive for serotype 2, 52 of which reacted with serotype 2 only. One additional and 3 additional reactions compromised the other two.

Additional phagocytic experiments indicated a preferential uptake by mouse peritoneal macrophages of the protein serotype 2 antigen, unopsonized, as opposed to a lipopolysaccharide serotype 5, and another protein, serotype 1, also unopsonized. It is possible, therefore, to speculate that differences in macrophage recognition and perhaps subsequent antigen presentation are involved in both preferential as well as specific responses.

Schroer, et al. referred to above, examined hybrid products from fused myeloma cells and spleen cells from mice with an X-linked immunodeficiency with the view to determining whether B cells capable of Tl-2 (thymus independent) responses existed. The antigen from this category was type III pneumococcal polysaccharide. Immunized F, male mice of the cross (CBA/N X Balb/c male) were used as the spleen cell donors. This ensured that the X chromosome involved in the defect was present in the hybrid cells. Eight hybridoma clones making antibody of the IgM class (radioimmunometric binding of iodinated anti-u), specific for type III pneumococcal polysaccharide were identified. Hence, it appeared that cells from mice bearing the X-linked genetic defect possess antigen-specific receptors through which activation occurs. The defect is, therefore, not a deletion of B cell subsets which respond to TI-2 antigens. This analysis of the defective immune response left the following possibilities open for further investigation: a defective intracellular mechanism for assembly and/or export of immunoglobulin in the X-linked defect, or a reconstitution event taking place with the myeloma parent genome. The latter cannot be ruled out since the NS-1 myeloma line used is a producer of Kappa chains.

Idiotypes of monoclonal antibodies of various immunoglobulin isotypes directed against the same antigen express individual, as well as cross-reactive, types. Since LPS produces a long-lasting and oscillatory primary and secondary splenic plaque-forming cell response in Balb/c mice, Hiernaux, et al., chose the LPS of *E. coli* 0113 as a likely system for the study of idiotypic expression and maturation during the course of LPS-specific immune response. It was found that idiotypic determinants of some MCAs are polymorphic. Shared idiotypes could be observed among various MCAs. There were three major cross-reactive idiotypes. During the primary immune response to LPS, the percentage of cross-reacting idiotype LPS-specific splenic plaque-forming cells (PFC) decreased from 82 percent on day 4 to 36 percent on day 30 and increased again to 63 percent on day 60. In hyperimmunized mice this percentage fluctuated from 20-50 percent, except on day 20 when 70 percent of LPS-specific PFC were cross-reactive idiotypes. The expression of cross-reactive idiotypes was independent of the major histocompatibility complex and IgC_H genes in the various strains of mice examined. A certain cross-reactive idiotype was expressed only after hyperimmunization. Additionally, because the cross-reactive idiotypes were expressed in all strains of mice examined, and on various classes of immunoglobulin, they are germ-line encoded.

Indirect Measurement of the Cellular Response to Bacteria with Monoclonal Antibodies

The characterization of cell types in mycobacterial infections which have long been thought to be accompanied by perturbations of the cellular immune response have proven enlightening with MCAs. Two approaches in humans have utilized the commercial availability of MCA to T cell subsets, to Ia antigens and to monocytes. Narayanan et al. used MCAs to define the nature of dermal infiltrates in skin biopsies obtained from untreated patients across the leprosy spectrum.

Using indirect immunofluorescence on cryostat sections, MCA defining T cells subsets and la-like antigens demonstrated ratios of OKT4$^+$/OKT8$^+$ cells of 1.2 to 5.0 in tuberculoid and 0.2 to 1.0 in lepromatous lesions. OKT4 and OKT8 are MCA to helper/inducer and suppressor/cytotoxic T cells respectively. OKT3 is a pan- T cell antibody. Most lymphocytes in the lesion reacted positively to this antibody as well as to MCA against Ia antigens. This indicated activated T cells. However, the numbers declined over the leprosy spectrum with a marked reduction in disseminated, multi-bacillary, lepromatous leprosy. Ia antigens on macrophages in the granulomas were expressed with or without intracellular acid fast bacilli.

Bach et al., assessed the immunological perturbations in leprosy on peripheral blood mononuclear cells in a similar manner. This was coupled with a measurement of the proliferative responses to mitogens, conconavalin A-induced suppressive activity in mixed lymphocytic culture and delayed type hypersensitivity. An imbalance between T cell subsets was thought to contribute to the occurrence of erythema nodosum leprosum reactions, believed to be an immune complex disease, in lepromatous patients. That is, there were decreased suppressor T cells with a simultaneous increase in helper T cells.

On an experimental basis, Mathew et al. immunized guinea pigs with *Mycobacterium bovis* (BCG) and with *Mycobacterium leprae*. Successful containment, killing, and degradation of BCG was characterized by the presence of epitheloid cells and fibrosis. The epitheloid cells had macrophage-specific antigen, but not Ia antigen as assessed with MCA. The macrophages in the *M. leprae*-induced granuloma expressed both macrophage-specific and la antigens. Epitheloid cells which were absent in the latter were considered, therefore, not to be involved in antigen presentation or other accessory functions where the la antigen is important.

Use of Transformed Cells

Few bacterial antigens have been examined using antibody derived from transformed host lymphocytes. However, monoclonality to a variety of antigens in Epstein-Barr transformed human lymphocytes has been demonstrated. This effectively defines the immune status of the individual and gives a profile of the B cell repertoire in response to a given antigen without deliberate immunization. It poses the possibility of combining the technology of transformed lymphocytes with hybridization techniques, given the availability of an appropriate plasmacytoma as described for human hybridomas secreting antibody to measles virus and, more specifically, Epstein-Barr transformed hybridomas secreting antitetanus toxoid antibodies.

Transformed Simian virus 40 spleen cells producing antibody have been defined experimentally in rabbits as early as 1974, but the appreciation of its unique potential has only recently come to the fore.

Comments on Therapeutic Uses of Monoclonal Antibodies in Bacterial Infections

Prophylactic therapy with MCAs is a most inviting application of hybridoma technology; particularly in infectious diseases which have a high treatment failure rate with antibotics. Combined therapy is also inviting. The obvious cautionary note is, of course, a thorough knowledge of the characteristics of the MCA(s) chosen. A precipitating antibody is of concern when thrombi are a possibility; a complement-fixing antibody may be undesirable where immune complexes are of concern. Cross-reacting antibodies to antigens which are highly conserved in nature, about which there is a plethora of literature respecting microbial antigens, must be cautiously viewed with respect to their adverse effect on host structures. One such instance is alluded to by Cross et al. The $\alpha 2 \rightarrow 8$ linked N-acetyl neuraminic acid in K1 polysaccharide of *E. coli* is also found in some gangliosides. The desired derivation of the antibody is from the same species to which it is administered to avoid serum sickness. This remains to be advanced in humans. However, examples such as the blocking activities of antibody for the toxin of *C. botulinum* to bind to synaptosomes make the therapeutic prospects for MCAs seem tangible.

19

Positive Selection of Monoclonal Antibody

The advent of media selective for particular biochemical phenotypes has provided the basis for much of the analytical molecular genetics taken for granted today. For example, assignment of many mammalian genes to specific chromosomes by correlation of gene expression with chromosome segregation in somatic cell hybrids was made possible by the availability of media which require expression of particular biochemical markers. Similarly, DNA mediated gene transfer (transfection) has permitted the introduction of new genetic information into cultured cells and has provided significant advances in the study of gene expression. Initially, use of transformation following transfection was limited to one of a handful of genes coding for enzymes whose expression could be selected for with appropriate selective growth media. Subsequently, however, it was realized that cells transformed by calcium phosphate/ DNA co precipitates and selected for one marker frequently contain and express other genes in the precipitate which were not selected for. This phenomenon is termed cotransformation. The exploitation of cotransformation has tremendously extended the usefulness of transfection, permitting as it does the introduction into cells of genetic material for which no selection exists. However, success of cotransformation usually requires that the nonselected gene be present in high abundance in the precipitated DNA, and for mammalian genes generally requires that a molecular clone of the gene be available. It is not easy, for example, to readily obtain cells transformed by genes present in single copy in mammalian DNA, because even though cotransformation and selection of a dominant marker gene greatly enriches a cell population for those cells expressing nonselected cellular genes, only a tiny fraction (perhaps 1 in 1000) of cells express any particular nonselected gene.

Development of means for isolating rare cells expressing specific mammalian genes after transfection with cellular DNA is particularly important since the ability to subsequently recognize and rescue the foreign DNA contained in recipient cells affords a means of obtaining molecular clones of the transforming gene. Several laboratories have used cotransformation in conjunction with screening methods to isolate the rare cells transformed with particular genes from cellular DNA. For example, immunofluorescent cell sorting permitted the isolation of mouse cells expressing human transferrin receptors and auto radiographic visualization of cell clones with bound ^{125}I-epidermal growth factor permitted isolation of mouse cells expressing human epidermal growth factor receptors.

While such screening procedures are quite useful, they are laborious and of limited power for isolating transformants which occur at very low frequency. Selective growth media, where available, are much preferable in this regard, and so we have attempted to develop a broadly applicable means for designing growth media conferring a significant growth advantage to cells bearing particular cell surface proteins. We have developed these selection techniques particularly for isolation of cells expressing novel cell surface receptor proteins after transfection. However, it is quite likely such techniques may also permit the isolation of cells which express increased quantities of proteins originally present. It has recently become evident that increased expression of proteins due to gene amplification is a rather general phenomenon and the selection techniques which we will describe should permit one to select for the products of such gene amplification events.

Polypeptides bound to cell surface receptors, and antibodies bound to some cell surface antigens are efficiently taken up by endocytotic mechanisms. Several laboratories have exploited this principal in the development of toxins conjugated to hormones and antibodies used to select *against* cells with particular receptors or antigens. We have reasoned that conjugation of hormones or antibodies to specific essential nutrients conversely might confer a selective growth advantage to cells expressing receptors for those hybrid ligands.

Cells growing in tissue culture are maintained in a synthetic medium containing a mixture of salts, sugars, amino acids and vitamins, which is usually supplemented with some biological fluid such as serum, amniotic fluid, lymph, milk, or colostrum. The role of these biological fluids apparently is to provide hormones, growth factors, and nutrients. For a variety of cell lines, completely defined growth media now have been described in which sera and other complex supplements have been replaced by highly purified components. In such media, insulin, selenium ion, and transferrin are almost universal requirements, and other hormones and growth factors are required for particular cell types. The rat pheochromocytoma PC12 cell line may be grown in a particularly simple serum-free medium consisting of RPMI 1640 supplemented only with transferrin, insulin and selenium. The dependency of PC12 cells on these supplements provided an opportunity to test the hypothesis that individual nutrients could be conjugated to specific antibodies and delivered to target cells in an antigen-dependent manner. Specifically, we have used antibodies containing selenomethionine to replace selenium oxide normally present in TIS medium, and we have used antibodies conjugated to ferritin to replace transferrin as a means of delivering nutritional iron. While we will demonstrate the efficacy of this approach only for antibody/cell surface antigen interactions, we believe that it is likely to be equally effective employing peptide hormone/receptor interactions.

Transferrin is listed as an essential component in many fully defined media. It is a major serum component and most cells express high affinity plasma membrane receptors for it, presumably mediating the uptake of iron from exogenous media. Since examination of coated vesicles obtained from human placental syncytiotrophoblast membrane revealed significant concentrations of both transferrin and IgG, it has been proposed that these ligands undergo the process of absorptive endocytosis. That is, ligands binding to specific cell surface receptors are localized in coated pits which rapidly enclose to form vesicles. Regardless of the precise mechanism, it is significant that these ligands were recovered from coated vesicles because this organelle is also a rich source of intracellular ferritin.

Ferritin is the iron containing form of apoferritin. Cells do not express plasma membrane receptors for ferritin, but it is thought to function intracellularly as a transporter or sequester of iron. Since ferritin has a physiological role for supply of iron, but does not specifically bind to the cell surface, it was chosen as the substitute iron source in a customized TIS medium.

C10-7 is a monoclonal antibody that recognizes a polypeptide antigen called PUNC present on PC12 cells, but not on a PC12 variant line,

NR18A. PUNC is not found on a variety of adult and neonate rat tissues but is present on neonate rate sympathetic neurons and adrenal chromaffin cells and therefore, apparently has specificity for neural crest derivative cells.

Ferritin conjugated to C10-7 IgG-2_a was substituted for transferrin in TIS medium. Similarly, seleno-IgG was prepared by growing the C10-7 hydridoma in medium in which selenomethionine was the only methionine source. Fe-IgG (C10-7) and Se-IgG (C10-7) were then tested individually and together for their ability to specifically support the proliferative growth of PC12 cells in the derined media. Since PC12 cells, but not the PC12 variant NR18A, express PUNC, it was expected that the C10-7-IgG conjugates would support the growth of PC12 but not NR18A cells. This was the case and the system was exploited for obtaining DNA mediated transfer of the PUNC gene.

Results

Components of Serum-Free (Fully Defined) Media

The success of the nutrient conjugate technique, in which cells growing in culture are made dependent on binding a specific antibody for their survival, depends on the availability of an appropriate fully defined growth medium. The means by which such a medium is developed is beyond the scope of this article, but has been fully described elsewhere. For the purposes of nutrient-conjugative selection cell growth in supplemented medium must be robust, and the requirement for the growth supplements to be replaced by conjugates must be absolute. In some cases, where a suitable serum-free medium cannot be found, prolonged culture of cells in a marginally suitable serum-free medium will select for a subpopulation of cells which grows well in that medium. For example we have recently isolated clonal variants of LTK⁻-B82 cells and human hepatoma HEP-G2 cells which grow more rapidly in TIS medium.

PC12 cells were originally isolated as a clonal derivative of a rat pheochromocytoma. These cells have been the focus of considerable attention because they possess nerve growth factor receptors and respond to nerve growth factor in a manner analogous to sympathetic neurons. In traditional serum-supplemented media, PC12 cells grow loosely attached to polystyrene tissue-culture dishes (eg. Corning flasks). Under conditions of serum-free culture however, the substratum surface must be prepared with a coating such as polymerized collagen or poly L lysine, as described in Materials and Methods. It should be noted that many cell lines do not require special substrata for attachment in serum-free media.

The serum-free medium we have utilized is a modification of a medium originally developed for rat neuroblastoma cells. It consists of RPMI 1640 supplemented with 5 mg/ml insulin, 50 mg/ml transferrin, and 4 ng/ml SeO_2. On a collagen substratum, PC12 cell growth rate in this medium is equivalent to that obtained with serum. The extent of PC12 cell dependency on transferrin and selenium is greater at lower cell densities, and the significance of this factor will be considered in the following sections.

Preparation and Purification of Ferritin-lgG Conjugate

We wished to determine whether ferritin conjugated to a monoclonal antibody that recognized PC12 cell surface antigens could be substituted for the essential component, transferrin, in serum-free medium. Nonspecific binding of ferritin to cell surfaces does occur and free ferritin, at high concentrations, will replace transferrin as an iron source. Nonconjugated immunoglobulin, if present, would compete with nutrient-conjugated immunoglobulin for antigen binding. Thus it was necessary to ensure that the ferritin-antibody conjugate was free of unconjugated ferritin or antibody. A detailed description of the preparation of the ferritin-antibody conjugate is presented in Materials and Methods. To summarize briefly, equine ferritin was activiated by reaction with glutaraldehyde, as described by Kishida et al, 1975. The unreacted glutaraldehyde was removed by rapid chromatography of the activiated ferritin on a column of Sephadex G25. C10-7 IgG, with a small amount of ^{35}S-methionine-labeled C10-7 IgG, was reacted with the activated ferritin. IgG-ferritin was separated from unconjugated IgG by chromatography on Sepharose 4B. IgG-ferritin then was separated from unconjugated ferritin by chromatography on columns of Protein A-Sepharose, as described in Materials and Methods.

The immunological specificity of the ferritin-conjugated IgG was tested by indirect immunofluorescence. PUNC antigen, recognized by C10-7 antibody, exists on PC12 cells in a punctate distribution. Although a limited degree of nonspecific binding of Fe-IgG (C10-7) was evident on NR18A cells, presumably mediated by the ferritin moiety, much greater binding was observed on PC12 cells. Note that PC12 and NR18A cells were allowed to attach to the same coverslip to permit direct comparisons of the level of immunofluorescence on the two cell types. The two cell types can easily be distinguished on the basis of their differing cell morphologies.

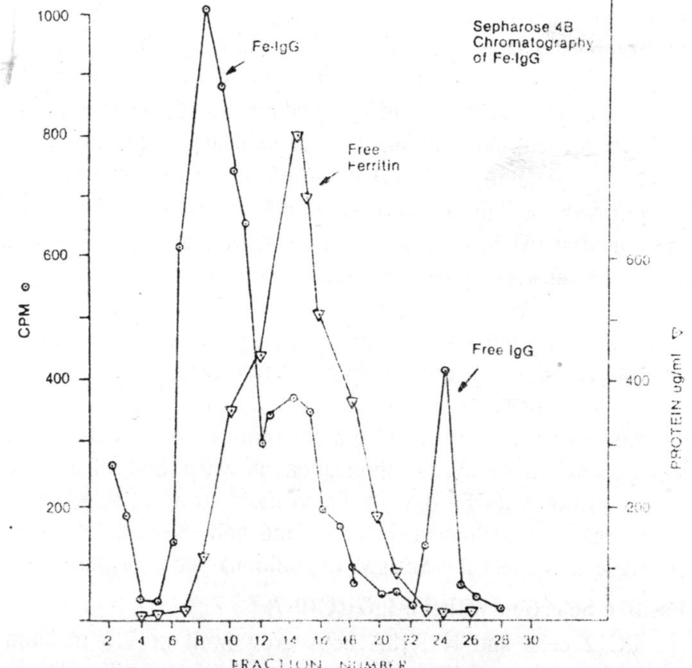

Fig. 19.1. Sephorose 4B chromatography of the ferritin-IgG conjugate.

Preparation of Se-IgG C10-7

As indicated, selenium is an essential nutrient component in many serum free media. Selenium is a cofactor for certain metabolic enzymes including glutathione reductase and occurs substituted in some nucleic acid molecules.

Optimal growth of PC12 cells in serum-free medium occurs with nanomolar concentrations of SeO_2. The extremely small amounts of this element required suggested that it would be ideally suited for the antibody-linked delivery system which obviously is limited in the amounts of material which it can deliver. Selenomethionine probably could be incorporated into IgG by chemical methods. However, for monoclonal antibodies, we have found it convenient to achieve this incorporation metabolically. Se-IgG (C10-7) was prepared by culturing C10-7 hybridoma cells in TIS medium (supplemented with 20 μM ethanolamine) in which methionine had been replaced by D,L-selenomethionine (15 μg/ml). To determine whether selenomethionine was being incorporated into immunoglobulin, in one experiment cells were cultured with 25 μCi/ml ^{75}Se-selenomethionine. Medium from the isotopically labeled cultures were chromatographed on Protein-A-Sepharose as described above, and analyzed by electrophoresis on a SDS polyacrylamide gel. Se-IgG is produced and the yield is comparable to that achieved under similar conditions using ^{35}S-methionine label.

For preparation of unlabeled Se-IgG, immunoglobulin was isolated from growth medium by adsorption on Protein-A-Sepharose, at pH 8, and eluted at pH 5, as described in Materials and Methods, and was dialyzed against phosphate buffered saline before use. Se-IgM; may be prepared similarly, except that Protein A chromatography cannot be used. For our purposes, it has been sufficient merely to dialyze hybridoma culture medium to remove free SeO_2 before use. Se-IgM prepared in this way is contaminated by the presence of small amounts of transferrin and insulin, which should not interfere in most applications. However, for applications in which Se-IgM is to be used in conjunction with Fe-IgM to obtain more stringent selective growth conditions by simultaneously delivering both Se and Fe, it may be desirable to subject the Se-IgM to additional steps of purification.

Positive Selection with Fe-IgG (C10-7)

PC12 cells and N R18A cells grow well in TIS medium. We tested the ability of Fe-IgG (C10-7) to deliver nutritional iron in a PUNC antigen-dependent manner, by comparing the ability of PC12 and NR18A cells to grow in medium with transferrin omitted (IS medium) and replaced by Fe-IgG (C10-7). Cells were plated in collagen-coated 16 mm wells at a density of 2×10^4 cells per well, in 1 ml of RPMI 1640 medium containing 10% horse serum and 5% newborn calf serum. The next day, the medium was removed, and the attached cells were washed twice with RPMI 1640 without serum. The medium was replaced by RPMI 1640, 5 μg/ml insulin and 4 ng/ml SeO_2 (IS medium), and the cells were cultured for 2 days. Cells typically continue to proliferate for several days after removal of transferrin, presumably utilizing internal pools of iron, and the purpose of this step is to deplete these stored reserves. Growth medium was then supplemented with either 50 μg/ml transferrin, free ferritin, IgG (C10-7) or Fe-IgG (C10-7). Cells were cultured for 8 days with 3 changes of medium, and the extent of cell growth was compared. Fe-IgG (C10-7) supported the growth of PC12 cells, and did so at extremely low concentrations. No growth stimulatory effect was observed for NR18A cells at these same concentrations. At high concentrations of ferritin (greater than 50 μg/ml) substantial cell growth was observed. Ferritin

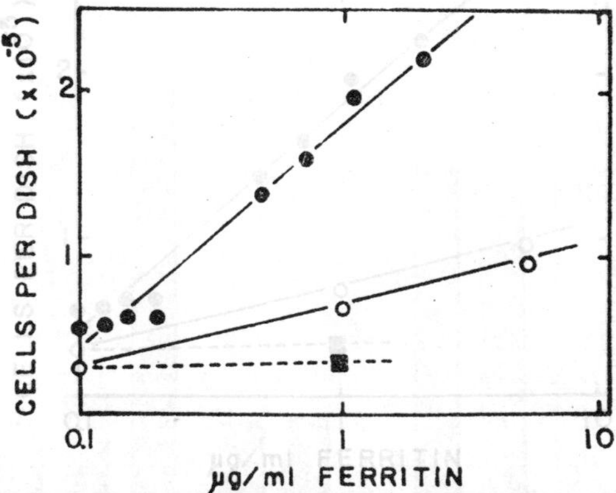

Fig. 19.2. Growth promoting activity of FeC10-7.

is well known to be endocytotically internalized in a nonspecific manner, and presumably supplies iron to cells by this means. Purified C10-7 IgG (non-conjugated) was not effective in supporting PC12 growth over a wide range of concentrations. However it must be noted that at the highest concentrations employed (50 μg/ml) significant cell growth was observed. This may be due to serum and ascites fluid contaminants (including transferrin) which may have copurified with the C10-7 IgG. This conclusion is supported by the observation that this growth-stimulatory effect is not antigen specific, since a similar effect was also seen on NR18A cells which lack PUNC antigen.

To confirm that the growth stimulatory effect of Fe-IgG (C10-7) was specifically mediated by the antibody, and was not due to some contaminant, we tested the ability of Protein A-Sepharose to absorb the active component. The eluate of the Protein A-Sepharose column no longer stimulated PC12 cell growth. Also shown in the histogram is an analysis of the ability of Fe-lgG (C10-7) to support the growth of NR18A cells. As anticipated, NR18A cells which lack PUNC antigen, grow poorly with Fe-IgG (C10-7).

Since the intended use of the nutrient-antibody selection system is to isolate rare antigen-expressing variants from populations of cells which lack the particular antigen, we tested the ability of Fe-IgG (C10-7) to accomplish this in a model system, produced by mixing small numbers of PC12 cells in varying proportions with NR18A cells. Mixtures of PC12 and NR18A cells were seeded on collagen coated wells at ratios of 1:1, 1:10, 1:100. 1:1000, and 1:10,000. Cultures were iron-deprived by growth for 3 days in IS medium, and then

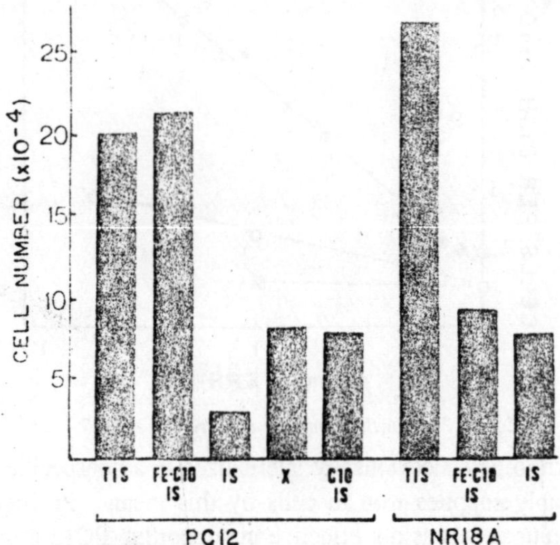

Fig. 19.3. Specific stimulation of PC12 growth by Fe-IgG (C10-7).

maintained with 1 μg/ml Fe-IgG (C10-7) for three weeks. PC12 cells could be recognized on the basis of their more rounded morphology. At the end of the selection period, homogeneous populations of PC12 cells were present in all wells, even in the wells initially consisting of 1:10,000 ratios of PC12 to NR18A cells.

Positive Selection with Se-IgG (C10-7)

The selenium requirement in serum free medium may also be exploited for nutrient-antibody selection. PC12 and NR18A cells were allowed to attach to collagen-coated wells as described previously, and were deprived of selenium for two days by growth in RPMI 1640 medium supplemented with insulin and transferrin, but with selenium omitted. This medium was then replaced with medium containing Se-IgG (C10-7) as the only source of selenium. As previously shown, free IgG is not growth-promoting at these concentrations. Se-IgG (C10-7) is effective however, and at lower concentrations than that seen with Fe-IgG (C10-7). At high concentrations of Se-IgG (C10-7) in fact, the conjugate appears toxic. Selenium toxicity is known to occur above optimal SeO_2 concentrations.

The ability of Se-IgG to replace the free selenium requirement of PC12 cells is not limited to PUNC antigen or to anti-PUNC IgG. This was demonstrated by preparing seleno-IgM monoclonal antibody, from hybridoma 5A7, metabolically in the same way in which Se-IgG,

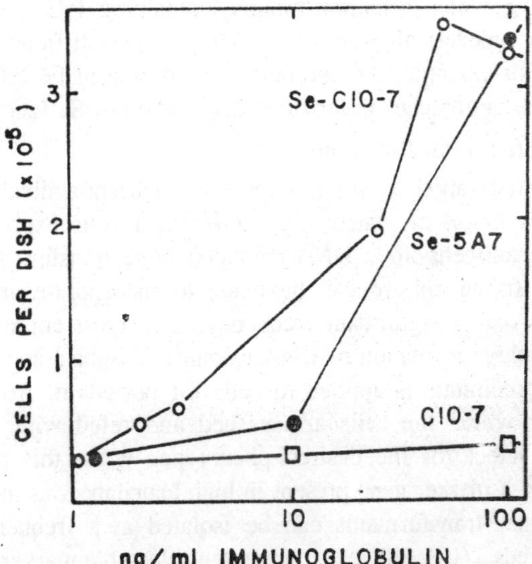

Fig. 19.4. Growth promoting activity of SeC10-7 and Se5A7.

(C10-7) was produced. Hybridoma 5A7 IgM, which recognizes a surface antigen distinct from C10-7 supports the growth of PC12 cells in the medium lacking any other selenium. The stringency of selection can be increased by depriving cultures of both iron and selenium, and using Fe-lgG (C10-7) and Se-lgG (C10-7) as the sole source of both nutrients. It should be noted however, that preparation of double

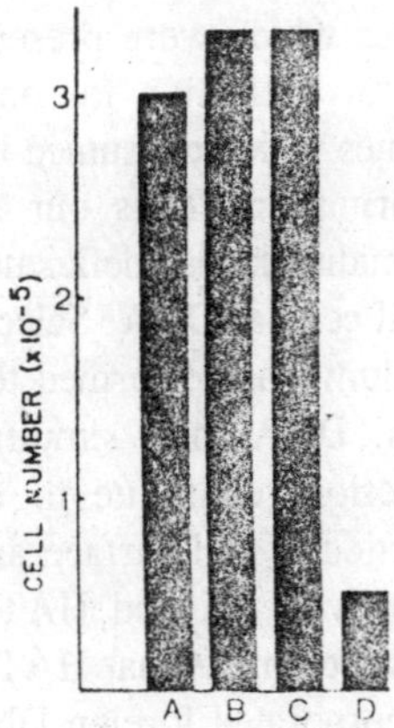

Fig. 19.5. Stringent selection by using C10-7 conjugates to provide nutritional iron and selenium simultaneously.

conjugates, in which selenomethionine containing IgG is coupled to ferritin are theoretically attractive but present difficult empirical problems. For example, the optimal concentration of Fe-IgG (C10-7), in our hands, approaches the toxic concentration of Se-IgG (C10-7).

DNA-Mediated Transformation

The primary application we propose for nutrient-antibody selection is for the isolation of genetically transformed cells expressing new cell surface antigens after DNA-mediated gene transfer. Fibroblasts growing in tissue culture can be made to incorporate and express foreign DNA at a significant frequency. The most commonly used protocol employs formation of DNA-calcium phosphate coprecipitates. The DNA precipitate is applied to cells for periods of from 2 to 24 hours, after which the cells are washed and refed with a medium which will select for the desired phenotype. When this protocol is applied using a marker gene present in high abundance, as in a purified plasmid clone, transformants can be isolated at a frequency which seldom exceeds $1/10^3$. However, when the selectable marker is present in low abundance, transformation frequencies are proportionately lower. For example, single copy genes in mammalian genomic DNA yield transformants at frequencies of $1/10^5$ or less. In populations of cells exposed to calcium phosphate precipitated DNA, it would appear that most cells do not stably retain foreign DNA, but those infrequent cells which do stably incorporate foreign DNA retain large quantities. This leads to the phenomenon called cotransformation, wherein cells transformed by a selectable marker present in high abundance frequently also stably express genes which were present in the DNA precipitate but were not selected for. Selection for one phenotype enriches the population for those clones which consumed other exogenous sequences at the time of transformation. Thus our strategy was to perform cotransformation of thymidine kinase-deficient (TK^-) cells with mixtures of plasmid TK DNA and cellular DNA. Selection of TK^+ transformants by culture in HAT medium was performed to isolate those cells which have integrated foreign DNA, and simultaneously we applied the nutrient-antibody selection technique to select clones which are expressed genes for particular cell surface antigens. The advantage of using contransformation with TK and HAT selection in conjunction with nutrient-antibody selections is that HAT selection, by eliminating cells which have not incorporated foreign DNA, reduces the degree of stringency required in the nutrient antibody selection, which might, by itself, be somewhat "leaky." It is possible that L cells, at a low frequency, might acquire the ability to survive in the selective medium by means other than transformation; for example, by activating an endogenous PUNC gene, or by inducing alternative nutrient transport systems. Reduction of the population of cells surviving nutrient-antibody selection, by prior HAT selection, should reduce the liklihood of such events occurring.

Transformation of LTK-B82 Cells

LTK-B82 cells, a common target for transformation, are mouse fibroblastic cells which are phenotypicaliy TK^- and do not revert to the TK^+ phenotype at a detectable frequency. In order that we might apply nutrient-antibody selection after transformation, it was necessary that we define a serum-free medium which would permit growth of L TK-B82 cells. In some cases, it might also be desirable to perform the transformation procedure in the absence of serum. Thus, we also tested whether transformation with calcium phosphate DNA coprecipitates could be accomplished in the absence of serum.

LTK-B82 cells will proliferate in a serum-free medium containing 5 μg/ml insulin, 50 μg/ml transferrin, 4 ng/ml SeO_2 and 30 ng/ml lutenizing hormone (LH). The doubling time of cells in this medium is 30 hours, which is somewhat slower than that obtained with serum added. LTK^- cells plated at low density form clones in this serum-free medium, but the efficiency of plating is only 20% that observed with medium containing serum. We tested whether LTK^- cells could be transformed in serum-free medium, using calcium phosphate precipitates of pTK, a pBR322derived plasmid containing the Herpes Simplex Virus type I tk gene. When cells were transfected in serum-free medium, cultured for one day in serum-supplemented medium, and then subjected to selection in serum-free medium supplemented with HAT, HAT-resistant transformants were obtained at a frequency approaching $1/10^4$.

Selection of PUNC-Positive L Cells Following Transfection with PC12 DNA

As a test of the utility of nutrient-antibody selection for isolation of genetic transformants we attempted to transfer the PC12 PUNC gene. PUNC antigen is expressed on PC12 cells, but not on L cells (data not shown) and therefore it should be possible to select L cell transformants expressing PUNC antigen.

L cells were prepared for transfection as before, and treated with a calcium phosphate precipitate of 1 μg/ml pTK and 40 μg/ml PC12 high molecular weight total genomic DNA, in medium containing serum, as described in Materials and Methods. After DMSO shock, and two days of recovery, the monolayers were washed twice with serum-free medium, cultured 2 days in serum-free medium supplemented with insulin, selenium, LH and HAT to deplete internal iron pools. Then the medium was replaced with medium additionally supplemented with 1 μg/ml Fe-IgG (C10-7), and cells were maintained in this medium for at least three weeks.

Colonies appeared at a frequency of between $1/10^5$ and $1/10^6$, which is substantially less than that seen for transformation with TK alone, without selection for PUNC expression as shown in Figure 8. Healthy clones were picked and grown to confluency in double selective medium. Growth was slow, and after confluent cultures were achieved, and frozen in liquid nitrogen, cultures were maintained in HAT-containing medium supplemented with serum. The removal of the nutrient antibody selection exposes the experiment to the risk that populations lacking PUNC might emerge and overgrow the cultures, if they have a growth advantage. However, cotransferred markers are often physically and genetically linked in the transformed cell and selection for one marker generally maintains the stable expression of the second unselected marker.

Transformants were examined for their expression of PUNC using an ELISA assay. This is a rapid assay, not requiring many cells, and many cultures can be conveniently tested at the same time. Briefly, about 10^5 cells were fixed to the bottoms of wells of 96 well costar trays and treated with dilutions of mouse anti-PUNC monoclonal antibody (or control antibody), were washed and treated with a peroxidase-linked goat anti-mouse IgG antibody, followed by addition of peroxidase substrates. Positive cultures were apparent by the rapid change in colour in the well. This assay revealed that 2 of the 8 putative cotransformants bound substantially more anti-PUNC antibody than did L cells. These two cultures SFM 24 and SFM 25, were expanded for further analysis. The other 6 transformants were not characterized further. They may express levels of PUNC too low to be detected by this assay, or they may survive in the selective medium for reasons unrelated to PUNC expression.

To examine the transformants more closely for expression of PUNC, cells were seeded on coverslips at low density and allowed to grow into small colonies of 20 to 40 cells. They were then fixed and prepared for indirect immunofluorescence. It showing that LTK^-B82

cells treated with C10-7 antibody do not fluoresce, while the transformant SFM 25 showed significantly more fluorescence than did the parent LTK^- cells.

Individual colonies of SFM 25 cells were either uniformly fluorescent, or uniformly lacking in fluorescence. The emergence of colonies which do not express PUNC antigen suggests that PUNC expression may be somewhat unstable in the absence of continued selection.

The presence of PUNC antigen on SFM 25 does not necessarily mean that the antigen is expressed from the rat gene. These cells might have activated an endogenous mouse gene which results in production of an immunochemically cross-reacting cell surface antigen. It will not be possible to distinguish between these two possibilities until we have characterized the PUNC antigen more fully biochemically.

Materials and Methods

Serum-Free Media

All serum-free media were prepared by supplementing RPMI 1640 medium with various nutrients in the form of sterile 100X concentrates, prepared as follows: *Transferrin*. Human transferrin (Sigma) was prepared at 5 mg/ml in phosphate buffered saline, sterilized by filtration, and stored frozen. *Selenium*. SeO_2 was dissolved in water at 4 mg/ml and then diluted 100 fold in water to yield a stock solution of 400 ng/ml, which was sterilized by filtration and stored frozen. Insulin. 50 mg insulin was dissolved in 100 ml of phosphate buffered saline, pH 5.0 and the pH was then adjusted to pH 6.0. The solution was sterilized by filtration and stored frozen.

Collagen and Polylysine Coated Culture Surfaces

To promote PC12 cell attachment in serum-free culture, plastic culture surfaces were coated with poly-L lysine or collagen. Poly L lysine (Sigma type B-1) was prepared at 100 μg/ml in distilled water, sterilized by filtration and stored under refrigeration. Culture vessels were incubated for 10 min with polylysine solution, washed with three changes of sterile water, and allowed to dry prior to use. Collagen was prepared from rat tail tendons and stored at 4°C. Approximately 20 μl of collagen solution was applied to 16 mm culture wells, and smeared to a uniform film with the end of a plastic pipet tip. The tray containing the wells was then placed in a chamber saturated with ammonium hydroxide vapor, and the collagen was allowed to polymerize for 5 min. Wells were then sterilized by exposure to UV light. Wells were washed twice with phosphate buffered saline, and stored overnight with a covering of 0.5 ml sterile phosphate buffered saline. More recently, the ammonium hydroxide treatment has been eliminated without consequence.

Ferritin Conjugation

The glutaraldehyde method of ferritin coupling to IgG was employed, as described by Kishida et al (1975). 1200-fold molar excess of glutaraldehyde per apoferritin amino group was used. 137 μl of 74 mg/ml horse ferritin in 1 ml of 0.1 M sodium phosphate buffer, pH 7.3 (10 mg ferritin total) was mixed with 2 ml of 30% glutaraldehyde. After 30 min at room temperature, aggregates were removed by centrifugation in a Beckman Microfuge for 4 min at 10,000 × g. The supernatent was subjected to chromatography on Sephadex G-25 to separate ferritin from unreacted glutaraldehyde, as follows: A column of Sephadex G-25 (coarse, Pharmacia) was prepared in a glass 25 ml pipet. The column was equilibrated with 0.1 M sodium phosphate, pH 7.3. The sample was applied and eluted at room temperature using the same buffer. Ferritin eluted in the void volume and was easily detected by its brown-yellow colour, which could be quantitatively monitored by absorbance at 435 nm. The activated ferritin fractions were immediately reacted with IgG, using a ferritin to IgG ratio of 4:1. We used ^{35}S-methionine-labeled C10-7 IgG as tracer. This was made by culturing the C10-7 hybridoma in the serum-free medium formulated with RPMI 1640 medium lacking methionine, and supplemented with 25 μCi/ml ^{35}S-methionine. Medium was collected after 2 days of culture and the IgG was isolated by chromatography on Protein A-Sepharose as described below. Activated ferritin (4 mls, 1 mg/ml) was added to 2 mls sodium phosphate buffer containing 1 mg/ml C10-7 IgG (prepared from mouse ascites fluid by ammonium sulfate precipitation), as well as small amounts of ^{35}S-methionine-labeled C10-7 IgG. After reaction for 3 hours at room temperature, and then an additional 22 hours at 4°C, the conjugate was separated from free IgG by chromatography on a column of Sepharose 48. A 25 cm long column of Sepharose 48 was prepared. The column was equilibrated with 0.1 M Tris-HCI, pH 7.5, 4°C, and the sample was eluted with the same buffer. Tris buffer was employed to block residual aldehyde groups on the ferritin.

Purification of IgG on Columns of Protein A-Sepharose

Mouse IgG subclasses 1, 2A and 2B can be absorbed onto Protein A-Sepharose and eluted at reduced pH, in high yield. Protein A-Sepharose 4B was purchased from Pharmacia. It was equilibrated with

10 mM sodium phosphate, pH 8.0, 0.1% sodium azide, and was packed into 1 ml tuberculin syringes adapted for use as columns. The column was stored at 4° until use. Sodium citrate buffers (0.1 M) of pH 3.0, 4.0, 5.0, and 6.0 were prepared, as well as 0.14 M sodium phosphate buffers with pH 6.5, 7.0 and 8.0. Fe-lgG (C10-7) in 0.1 M sodium phosphate, pH 7.3, was adjusted to pH 8.0 and applied to the column. The column was sequentially eluted with 2 ml aliquots of the various buffers from pH 7.0 to 3.0. Elution of IgG was followed by liquid scintillation counting of the ^{35}S-methionine label, and elution of ferritin was monitored by absorbance at 435 nm. Immunospecificity of eluted IgG conjugate was determined by indirect immunofluorescence with PC12 cells. Fe-lgG (C10-7) eluted at pH 6.0. It was dialyzed versus phosphate buffered saline and was stored frozen.

DNA Mediated Transfer

Target cells are seeded at low density (about 4×10^5 cells per 25 cm^2) one day prior to transfection. The cells are then treated with a calcium phosphate-DNA coprecipitate as originally described by Graham and van der Eb (1973) as modified by Wigler et al (1978). RPMI 1640 medium was replaced by DMEM medium prior to transfection because the higher phosphate content of RPMI 1640 medium leads to excessive formation of calcium phosphate percipitates. DNA is prepared in 1 mm Tris-CI, 0.1mMEDTA, pH 7.9 and is used at 40 μg/ml. $CaCl_2$ is added to a final concentration of 250 mM, and this solution is dripped slowly into 280 mm NaCl, 50 mm HEPES, 1.5 mm sodium phosphate, pH 7.12, with constant agitation. After 30 minutes at room temperature, a fine precipitate appears at which point the solution is diluted into 10 mls of serum-supplemented DMEM medium. This is added directly to the cell culture. After 4 hours of incubation at 37° (in an atmosphere of 10% CO_2) cell monolayers are washed with fresh medium and incubated 3 min. at room temperature with growth medium containing 4% dimethyl sulfoxide. After two minutes this is replaced by growth medium. The next day cells are subjected to selective growth conditions (e.g. HAT medium) and transformants are generally apparent within two weeks.

INDEX

S

T

V